Geospatial Data Science Quick Start Guide

Effective techniques for performing smarter geospatial analysis using location intelligence

Abdishakur Hassan
Jayakrishnan Vijayaraghavan

BIRMINGHAM - MUMBAI

Geospatial Data Science Quick Start Guide

Commissioning Editor: Amey Varangaonkar
Acquisition Editor: Devika Battike
Content Development Editor: Roshan Kumar
Technical Editor: Snehal Dalmet
Copy Editor: Safis Editing
Project Coordinator: Namrata Swetta
Proofreader: Safis Editing
Indexer: Manju Arasan
Graphics: Alishon Mendonsa
Production Coordinator: Alishon Mendonsa

First published: May 2019

Production reference:1300519

Published by Packt Publishing Ltd.
Livery Place
35 Livery Street
Birmingham
B3 2PB, UK.

ISBN 978-1-78980-941-1

www.packtpub.com

To my wife, Sirad, for being my partner in life. To my children, Mohamed, Manal, and Almas, for always being a source of inspiration. To my mom, Mako, for raising us in difficult times, and finally, to my father, Awil, for being a guiding mentor and source of support.

– Abdishakur Hassan

To the three most beautiful women in my life: my mom, Vijayakumari, whose kindness is eternal and unfathomable; my wife, Athulya, who is amazing in every way imaginable; and my sister, Sujatha, who is always, always, an inspiration.

– JK

`mapt.io`

Mapt is an online digital library that gives you full access to over 5,000 books and videos, as well as industry leading tools to help you plan your personal development and advance your career. For more information, please visit our website.

Why subscribe?

- Spend less time learning and more time coding with practical eBooks and Videos from over 4,000 industry professionals

- Improve your learning with Skill Plans built especially for you

- Get a free eBook or video every month

- Mapt is fully searchable

- Copy and paste, print, and bookmark content

Packt.com

Did you know that Packt offers eBook versions of every book published, with PDF and ePub files available? You can upgrade to the eBook version at `www.packt.com` and as a print book customer, you are entitled to a discount on the eBook copy. Get in touch with us at `customercare@packtpub.com` for more details.

At `www.packt.com`, you can also read a collection of free technical articles, sign up for a range of free newsletters, and receive exclusive discounts and offers on Packt books and eBooks.

Contributors

About the authors

Abdishakur Hassan is a geographic information systems (GIS) expert and consultant with over 5 years of experience working with UN-Habitat. He holds an MSc in geoinformation science and earth observations. During his tenure as a GIS expert, Abdishakur has developed fully fledged GIS applications in the urban planning and land administration domains. He is interested in all things related to geospatial data science.

I would like to thank my wife, Sirad, who raises our three beautiful children, always prioritizing our family over her professional career, and succeeding in managing and balancing family life with her professional career. My children, Mohamed, Manal, and Almas, have also been an inspiration during writing of this book. I would also like to thank Roshan and Snehal and the editing team at Packt for always being constructive during this period.

Jayakrishnan Vijayaraghavan is a geospatial data scientist, innovator, and author of a book titled *ArcGIS for JavaScript developers*. He currently resides in the San Francisco Bay Area and has over 8 years of work experience. He has built patented technologies and products in the geospatial domain and has coached teams on leveraging mapping and spatial analytics tools for solving pertinent business problems. He is adept at computational geometry, especially in graph networks and in geospatial inferencing. He is a DAAD scholar and a winner of the UN-Habitat special jury award. He is keen on developing intelligent and ubiquitous mapping systems by integrating ML and DL techniques with GIS. He is also a novelist and a certified UAV pilot.

I would like to thank the Packt publishing team; the editors; my coauthor, Abdishakur; and the reviewers for helping me to bring this book, which I was so passionate about, to fruition. Also, a special shout-out to Roshan and Snehal, whose tireless support is highly appreciated.

About the reviewers

Jyoti Rohodia completed her bachelor of technology degree in geoinformatics at the University of Petroleum and Energy Studies. Her bachelor's curriculum included specializations in applied petroleum geology and geophysics and remote sensing. Currently, she is working as a senior GIS analyst at CyberTech Systems and Software Ltd, and provides GIS solutions to USA-based clients of ESRI Inc. She has extensive GIS proficiency working with the following ESRI products: ArcGIS Desktop, ArcGIS Pro, ArcGIS Online, and ArcGIS Enterprise. Additionally, she has sound knowledge of automating GIS workflows using the ArcPy module and ArcGIS API for Python.

Shital Dhakal is a seasoned GIS professional with over seven years' experience in the field of GIS and remote sensing. He has acquired industry and research experience in North America, Europe, and Asia. Currently, he works at a San Francisco Bay Area-based start-up and helps local government to implement enterprise GIS strategies. He is a certified GIS Professional (GISP) and has an MSc from Boise State University, Idaho. When he is not playing with spatial data, writing blogs, or making maps, he can be found hiking in the Sierra Nevada and, occasionally, in the Himalayas.

Packt is searching for authors like you

If you're interested in becoming an author for Packt, please visit authors.packtpub.com and apply today. We have worked with thousands of developers and tech professionals, just like you, to help them share their insight with the global tech community. You can make a general application, apply for a specific hot topic that we are recruiting an author for, or submit your own idea.

Table of Contents

Preface

Geospatial Data Science Quick Start Guide is all about providing a segue for data scientists into mapping technologies, and for GIS scientific researchers into data science and machine learning techniques. Geospatial data science is the core technology used at companies including Uber, Google Maps, Apple Maps, Foursquare, and by real estate search apps such as realtor.com. An integral reason for the enormous success of these companies can be traced to their early successes in effectively identifying and leveraging the power of location data in the context of data science. Location data is a pervasive component of data repositories of any company in Silicon Valley, and the actual benefits of this data are yet to tapped due to a lack of expertise in this area. *Geospatial Data Science Quick Start Guide* will try to bridge this gap by equipping the reader with important skills for handling location data and effectively synergizing location data with machine learning techniques.

Who this book is for

Data scientists who would like to leverage location-based data and want to use location-based intelligence in the models will benefit from this book considerably. This book is also for GIS Pythonistas who would like to sneak into the magical world of data science and make their analyses more powerful than ever before.

What this book covers

Chapter 1, *Introducing to Location Intelligence*, introduces the reader to location data and location data intelligence. It provides real-world examples of location data intelligence. This serves as a basic introduction, and has one section of code working on real-world examples of location data intelligence.

Chapter 2, *Consuming Location Data Like a Data Scientist*, covers machine learning models for predicting the trip time of taxi journeys based on location factors and other attributes. In this chapter, we will discuss how to leverage spatial data masquerading as tabular data, and apply machine learning techniques to it as any data scientist would.

Chapter 3, *Performing Spatial Operations Like a Pro*, lays the groundwork for dealing with geospatial data. In this chapter, we cover the basics of GeoDataFrame, coordinate systems and projections, as well as spatial operations such as buffer analysis and spatial joins. We look into foundational, as well as advanced, location data intelligence techniques using the Foursquare dataset.

Chapter 4, *Making Sense of Humongous Location Datasets*, explores ways to aggregate location data into meaningful chunks using machine learning clustering techniques, and deriving more value from it. This chapter further introduces the reader to topics such as spatial autocorrelation, and both global and local spatial autocorrelation are explored and discussed.

Chapter 5, *Nudging Check-Ins with Geofences*, introduces to the reader to geofencing, which is a popular tool that sees use in contexts from businesses to conservation work efforts. Geofencing refers to abstract fences that are created around a location, such that an alert or notification system can notify the relevant party should an event happens at or within the fence. The event can be something as simple as a customer entering the vicinity of a business location, or customers moving within range of a cell-phone tower – the applications are unlimited. This chapter moves onto methods for quickly building and deploying a geofencing system using Python.

Chapter 6, *Let's Build a Routing Engine*, teaches the reader about navigation and routing, which are indispensable features in today's apps. Be it navigational apps, food delivery apps, or a courier delivery app; routing is a key component to the effective delivery of their services. And most of these apps rely on a few key third-party APIs, such as Google Maps APIs, to provide such services. Do we really need to pay these vendors to build a simple routing app? This topic explains how to build our own routing and navigation solutions using open source data and Python libraries that implement graph algorithms.

Chapter 7, *Getting Location Recommender Systems*, is the final chapter of the book and discusses the process of building recommender systems that do not only apply to products on Amazon or movies on Netflix, but also locations. This chapter leverages popular recommender system techniques on offer, including collaborative filtering methods and location-based recommenders. We will use these recommender systems to build a restaurant venue recommendation system.

To get the most out of this book

This following is a short list of requirements to successfully explore this book:

- **A computer with a browser**: This book primarily uses Google Colaboratory Jupyter Notebooks. To be able to use the code effectively, you need a browser and internet access.
- **Basic Python programming skills**: The reader should have at least introductory knowledge of data types and functions in the Python language. GIS skills will be helpful, but this is not required.

Download the example code files

You can download the example code files for this book from your account at `www.packt.com`. If you purchased this book elsewhere, you can visit `www.packt.com/support` and register to have the files emailed directly to you.

You can download the code files by following these steps:

1. Log in or register at `www.packt.com`.
2. Select the **SUPPORT** tab.
3. Click on **Code Downloads & Errata**.
4. Enter the name of the book in the **Search** box and follow the onscreen instructions.

Once the file is downloaded, please make sure that you unzip or extract the folder using the latest version of:

- WinRAR/7-Zip for Windows
- Zipeg/iZip/UnRarX for Mac
- 7-Zip/PeaZip for Linux

The code bundle for the book is also hosted on GitHub at `https://github.com/PacktPublishing/Geospatial-Data-Science-Quick-Start-Guide`. In case there's an update to the code, it will be updated on the existing GitHub repository.

We also have other code bundles from our rich catalog of books and videos available at `https://github.com/PacktPublishing/`. Check them out!

Download the color images

We also provide a PDF file that has color images of the screenshots/diagrams used in this book. You can download it here: `http://www.packtpub.com/sites/default/files/downloads/9781789809411_ColorImages.pdf`.

Conventions used

There are a number of text conventions used throughout this book.

`CodeInText`: Indicates code words in text, database table names, folder names, filenames, file extensions, pathnames, dummy URLs, user input, and Twitter handles. Here is an example: "Mount the downloaded `WebStorm-10*.dmg` disk image file as another disk in your system."

A block of code is set as follows:

```
from shapely.geometry import Point
for lat, lon in zip(nyc['Latitude'][:5], nyc['Longtitude'][:5]):
    geometry = Point(lat, lon)
     print(geometry)
```

When we wish to draw your attention to a particular part of a code block, the relevant lines or items are set in bold:

```
ax.set_xticklabels([])
ax.set_yticklabels([])
ax.set_title('Foursquare Points')
plt.show()
```

Any command-line input or output is written as follows:

```
!pip install networkx
```

Bold: Indicates a new term, an important word, or words that you see onscreen. For example, words in menus or dialog boxes appear in the text like this. Here is an example: "Select **System info** from the **Administration** panel."

 Warnings or important notes appear like this.

 Tips and tricks appear like this.

Get in touch

Feedback from our readers is always welcome.

General feedback: If you have questions about any aspect of this book, mention the book title in the subject of your message and email us at `customercare@packtpub.com`.

Errata: Although we have taken every care to ensure the accuracy of our content, mistakes do happen. If you have found a mistake in this book, we would be grateful if you would report this to us. Please visit `www.packt.com/submit-errata`, selecting your book, clicking on the Errata Submission Form link, and entering the details.

Piracy: If you come across any illegal copies of our works in any form on the Internet, we would be grateful if you would provide us with the location address or website name. Please contact us at `copyright@packt.com` with a link to the material.

If you are interested in becoming an author: If there is a topic that you have expertise in and you are interested in either writing or contributing to a book, please visit `authors.packtpub.com`.

Reviews

Please leave a review. Once you have read and used this book, why not leave a review on the site that you purchased it from? Potential readers can then see and use your unbiased opinion to make purchase decisions, we at Packt can understand what you think about our products, and our authors can see your feedback on their book. Thank you!

For more information about Packt, please visit `packt.com`.

1
Introducing Location Intelligence

"Everything that happens, happens somewhere."
  - The first law of geography by Waldo Tobler

Location data is data with a geographic dimension. Location data is everywhere as all actions that occur in or near the Earth's surface happen to use geographic aspects. It is generally referred to as any data with coordinates (latitude, longitude, and sometimes altitude) but also encompasses different aggregated geographic units, including addresses, zip codes, landmarks, districts, cities, regions, and much more.

Location intelligence, on the other hand, is the process of turning geographic (spatial) data into insights and business outcomes. Any data with a geographical position, either implicitly or explicitly, requires location-aware preprocessing methods, visualization, as well as analytical methods to derive insights from it. Thus, location intelligence applications can reveal hidden patterns of spatial relationships that cannot be derived through other normal means. It leads to better decision making on spatial problems, where things happen, why they happen in some places, and the spatial trends in time-series analysis. Understanding the location dimension of today's challenges in, industrial, retail, agricultural, climate, and environment, can lead to a better understanding of why economic, social, and environmental activities tend to locate where they are.

In this chapter, we give an overview of location data and location data intelligence. Here, we briefly introduce different location data types and location data intelligence applications and examples. We cover how to identify location data from publicly available open datasets. We briefly discuss and highlight the difference between location data and other non-geographic data. At the end of this chapter, we explore how location data fits into data science and what opportunities and challenges bring location data into the interdisciplinarity of data science.

We will specifically focus on the following topics:

- Location data
- Location data intelligence
- Location data and data science
- A primer on Google Colab and Jupyter Notebooks

Location data

What is location data and why is it different than other data formats? It is quite common to see phrases such as *spatial data is special* or another more popular adage, *80% of data is geographic*. While these are not easily provable, we tend to witness an increased amount of location data. From geotagged images, text, and sensor data, location data is ubiquitous and the world is *datafied*. In this connected and data-driven driven era, we generate, keep track of, and store huge mounts of data every day. Think of the number of tweets, Instagram images, bank transactions, searches on the web, and routing requests from APIs. We collect more data than at any other period of time in the past, and thus the big data revolution. Many of the datasets collected have an inherent location dimension but are often hidden within the data and not utilized fully.

Understanding location data from various perspectives

We can examine location data from different perspectives: business, technical, and data perspectives.

From a business perspective

From a business perspective, the value of maps and location data is crucial in many business applications. A quick look at big companies such as Google, Apple, Microsoft, and Nokia shows that each of these companies has their own location and mapping services and products.

Think about how often you use Google Maps API's location service through your phone. This also highlights the importance of location data as all these companies would not go to such lengths to have their own in-house location data production if it was not necessary. Business applications in location data include not only individual uses of location data but also innovative applications spanning from individualized marketing, autonomous vehicles, logistics, and transportation to healthcare.

From a technical perspective

The technical perspective of location data indicates that it entails both opportunities as well as challenges. Location data, in contrast to other data, has a topology, which holds the relationships between geometry (points, lines, and polygons) and geographic features that they represent. In the case of conventional data, we store data into tables or a **Relational Database Management System** (**RDBMS**). However, spatial relations and topology require us to store the geometry of objects.

Due to the nature of location data, which is derived from **Tobler's first law** of geography, *Everything is related to everything else, but near things are more related than distant things.* The essence of this law entails also the presence of strong autocorrelation and interdependency in continuous near locations, which is not necessarily present in conventional data (non-spatial attributes).

From a data perspective

Having looked into the nature of location data from a technical perspective, let's also examine it from a data perspective. How is location data different than other data? In location data, we use geographic coordinates (2D) to represent the world (3D).

For example, **Digital Elevation Models (DEMs)** are used to represent heights and terrain surface. The first law of geography applies here as well. At a certain point of time, a particular terrain is very likely to have the same height with its relatively close surrounding, while we can expect a difference based on elevation in two areas distant from each other. As mentioned earlier, spatial autocorrelation in location data is assumed to be present in spatial data, while in other types of data, such as the statistical analysis of conventional data, we assume the independence of data points. That means location data can be categorized as **stochastic**, while other data is **probabilistic**.

Another complication in location data also arises from what we call **Modifiable Area Unit Problem (MAUP)**, which arises from different aggregated units that produce different results. An example of this is poverty or crime estimates and aggregations. For example, areas of high poverty rates could be overestimated or underestimated depending on the boundaries of measured areas. By moving into different aggregations (that is, zip code, neighborhood, or district level), which can create different impressions and patterns created by the different scales and aggregations.

Types of location data

Geographic data types can be divided into two broad categories:

- **Vector data**: This is represented as points, lines, or polygons. The data is likely created by digitizing it and storing information in longitude and latitude. This type of data is useful for storing data that has discrete and distinct boundaries such as borders, land parcels, streets, and points of interest.
- **Raster data**: This stores information in cells and therefore is suitable for storing data that is continuous, such as satellite images, elevation models, and other aerial photographs.

Location data intelligence

Every industry uses **location intelligence**. It helps industries understand what their customers are doing, where their customers are based, what the geographic environment of their customers is, and what their interests are. Location intelligence is normally defined as using location data with other attributes to add context and derive useful information, services, and products that help organizations make effective and efficient decisions. The information derived through location intelligence can have a business and economic insights as well as environmental and social insights.

Application of location data intelligence

To illustrate how location intelligence is applied in a real-world application, we will take as an example **Foursquare** check-ins. Foursquare initially started in 2009 as a social platform to collect user check-ins and provide guides and search-results for its users to recommend places to visit near the user's current location. However, recently, Foursquare repositioned itself as a less social platform to a location intelligence company. The company describes itself as a *"technology company that uses location intelligence to build meaningful consumer experiences and business solutions"* and claims the following:

> *"If it tells you where, it's probably built on Foursquare."*

In its anonymized and aggregated trends of check-ins in physical brand locations, Foursquare provides insights and metrics that were not easily available before. Take, for example, the loyalty of customers, frequency of their visits, brand losses, and profits. This allows analysts and brands to understand their customers, reveal demographic insights and track patterns of customers, and look into and understand competition brands. To illustrate how powerful location intelligence is, let's explore a subset of Foursquare data in NYC. We will use this dataset later in `Chapter 3`, *Performing Spatial Operations Like a Pro*, but for now let's look into what it consists and how location intelligence is derived from it.

The NYC Foursquare check-in dataset has 10 months' worth of data spanning from April 12, 2012 to February 16, 2013.

Source: NYC Foursquare Check-in dataset first appeared in *Fine-Grained Preference-Aware Location Search Leveraging Crowdsourced Digital Footprints from LBSN, Dingqi Yang, Daqing Zhang, Zhiyong Yu,* and *Zhiwen Yu,* proceedings of the 2013 *ACM International Joint Conference on Pervasive and Ubiquitous Computing (UbiComp* 2013), September 8 to 12, 2013, in Zurich, Switzerland.

The following table shows the first five rows of the data and consists of eight columns with a unique `UserID` and `VenueID`. Both of these features are anonymized for privacy issues; `VenueCategoryID` and `VenueCategoryName` indicate aggregated types of business. Here, we have more than 250 business types, including a medical center, arts store, burger joint, hardware store, and so on; `Latitude` and `Longitude` columns store the geographic coordinates of the venues.

The last two columns indicate the time of the check-in:

	UserID	VenueID	VenueCategoryID	VenueCategoryName	Latitude	Longtitude	Timezone	UTCtime
0	470	49bbd6c0f964a520f4531fe3	4bf58dd8d48988d127951735	Arts & Crafts Store	40.719810	-74.002581	-240	Tue Apr 03 18:00:09 +0000 2012
1	979	4a43c0aef964a520c6a61fe3	4bf58dd8d48988d1df941735	Bridge	40.606800	-74.044170	-240	Tue Apr 03 18:00:25 +0000 2012
2	69	4c5cc7b485a1e21e00d35711	4bf58dd8d48988d103941735	Home (private)	40.716162	-73.883070	-240	Tue Apr 03 18:02:24 +0000 2012
3	395	4bc7086715a7ef3bef9878da	4bf58dd8d48988d104941735	Medical Center	40.745164	-73.982519	-240	Tue Apr 03 18:02:41 +0000 2012
4	87	4cf2c5321d18a143951b5cec	4bf58dd8d48988d1cb941735	Food Truck	40.740104	-73.989658	-240	Tue Apr 03 18:03:00 +0000 2012

Foursquare data: first five rows

Here, we have the first five rows of the Foursquare data. In this chapter, we will only look at the data from a wider perspective. The code for this chapter is available, but you do not need to understand it right now. We will come to learn the details of reading and processing location data with Python in the next chapters.

So, what kind of location intelligence can be derived from this type of data? We will cover this from two broad perspectives: the user/customer perspective and the venue/business perspective.

User or customer perspective

Here we will get a clear idea from a customer perspective. Often the following questions will come into picture:

Where does customer X spend his/her time? What does this place offer? How often does he/she visit these places? When does he/she visit these places?

The code of this section is available in the accompanying Jupyter Notebook. You do not need to understand all the code right now, as it serves to give you the bigger picture of location data intelligence. Feel free to consult Jupyter Notebook of this chapter if you want to run the code and experiment.

Let's take an example for the `UserID = 395` from the fourth row in the preceding table. This particular user has made `106` check-ins in total during this period of the dataset visiting `36` unique venues in NY (visualized as the map as follows):

User 395: Venues visited in NY

We can also look at what type of venues this particular user has visited. In this case, this user has visited frequently an office, a residential building, and a gym, in NYC. Other less-visited venues include an airport, outdoors, a medical center, and many others, as you can see from the following graph:

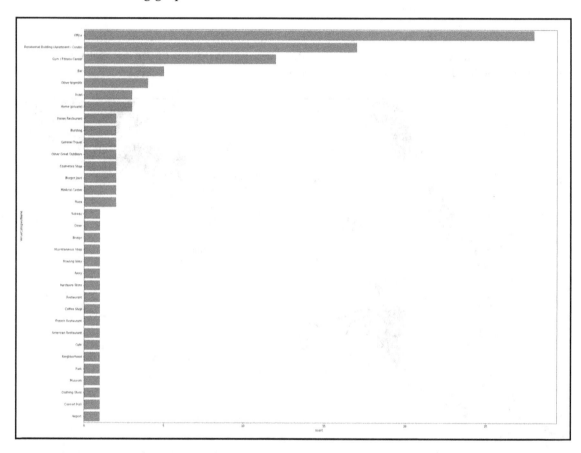

User 395: Check-ins plot

The user perspective can elicit many aspects related to the frequency of visits, preferences, and activities of the user that can guide location intelligence and decision making. Privacy issues in location data are very sensitive and require diligence. In this case, although it is anonymized data, it still reveals patterns and other useful information as we have shown. Now let's also look from the business perspective in the following section.

Venue or business perspective

Here we will get a clear idea from a business perspective. Often the following questions will come into picture:

How many customers does venue X Receive per day? What about per hour? What is the pattern? Can we estimate business value based on the check-ins?

We will use a gym venue as an example here, with `VenueID` = `4aca718ff964a520f6c120e3`. For this dataset, this gym has `118` check-ins. Although the data is small and cannot be generalizable in this particular `VenueID`, imagine it has enough data for a longer period of time. We can estimate the peak times of this gym as the following graph shows. There is a peak of check-ins at 14:00 and at 20:00:

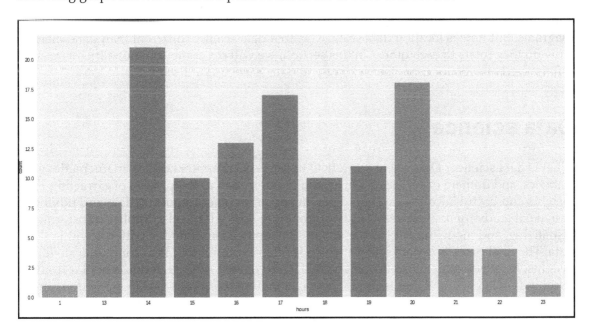

Gym visit check-ins: per hour

This kind of business perspective analysis helps both decision makers and competitors to gain an insight into businesses. This is only an individual business example, but this can simply be extended to businesses in this dataset and look further into it. In fact, Foursquare predicted Chipotle's sales (link available in the information box), a Mexican grill, to drop 30% during the months of 2016 before the company announced its loss.

Foursquare predicted Chipotle's Sales Will Plummet 30%: `http://fortune.com/2016/04/15/chipotle-foursquare-swarm/`.

Let's now look at how location data science is different than data science in the next section.

Location data science versus data science

Now that we have learned that location data is beyond mapping, and specifically is manipulation and processing of geographic data and applying analytical methods, we will move into the interdisciplinarity of location data science. We have also studied location data intelligence and how insights are derived from location data by illustrating this with diagrams. But how is location data science (spatial data science) different than data science? How do they relate to each other? In this section, we will cover the commonalities as well as differences between location data science and data science as a discipline.

Data science

What is **data science**? Data science as a field consists of computer science, mathematics and statistics, and domain expertise and is generally referred to as the process of extracting insights and useful information from data. Mostly, it involves importing data and tidying it to make it ready for analysis. An iterative process of data science also implies transforming, visualizing, and modeling data to understand phenomena and hidden patterns within the data. The final process in data science which is often explored less, is to communicate the insights. Now you may realize from what we have covered so far that this is not far from location intelligence, and that is right. The location dimension is critical in many domains and applications with data science. Next, let's look at what spatial data science.

Location (spatial) data science

Adding location data and the underlying spatial science entails additional challenges and opportunities. It will form a combination of the interdisciplinary field consisting of computer science, mathematics and statistics, domain expertise, and spatial science. This does not only indicate the addition of spatial science but also whole new concepts, theories, and the application of spatial and location analysis, including spatial patterns, location clusters, hot spots, location optimization, and decision-making, as well as spatial autocorrelation and spatial exploratory data analysis. For example, in data science, histograms and scatter plots are used for data distributions analysis, but this won't help with location data analysis, as it requires specific methods, such as spatial autocorrelation and spatial distribution to get location insights.

To get the reader up and running quickly and without burdening the local setup of Python environments, we will use Google Colab Jupyter Notebooks in this book. In the next section, we will cover a primer on how to use Google Colab and Jupyter Notebooks.

A primer on Google Colaboratory and Jupyter Notebooks

Jupyter Notebooks have become the favorite tool for data scientists, as they are flexible and combine code, computational output/multimedia, as well as comments. It is free, open source, and provides computational capabilities and interactive web-based notebooks. Anaconda distributions make the installation process easy if you want to install Jupyter Notebook on your local machine. The official Anaconda documents to install Jupyter Notebooks and Python is easy to follow and intuitive, so feel free to follow the instructions if you would like to work on your local machine.

However, we will use **Google Colab**, which is a free Jupyter Notebook environment that requires no installation or setup and runs entirely in the cloud, just like using Google Docs or Google Sheets. Google Colab enables you to write code, run the code, and share it. You just need to have a working Gmail to save and access Google Colab Jupyter Notebooks. In heavy computational tasks, such as machine learning or deep learning with big data, Google Colab allows you to use its **Graphics Processing Unit** (**GPU**) or **Tensor Processing Unit** (**TPU**) for free.

Google Colab interface is shown as follows. In the upper part, you have the main menu. The right part is where we can write our code and comments:

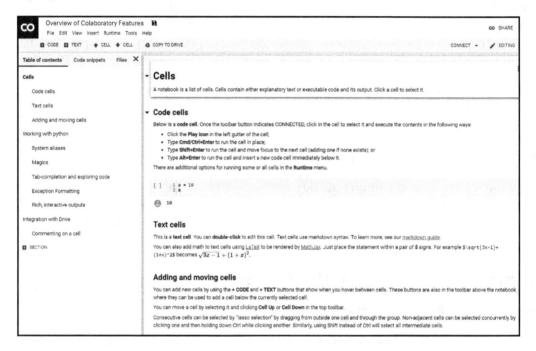

Google Colab

You can open Google Colab from this URL: `https://colab.research.google.com`. There are two main types of cells: code and text. With a code cell, you can write your code and execute it, while a text cell allows you to write down your text with a markdown. Here, you can have different text types, including several heading levels as well as a bulleted and a numbered list. To execute a cell, you can either use a *Ctrl + Enter* shortcut or press the **Run** button (small triangle) next to the cell.

We will learn Google by using it as our coding platform for this book. If you are new to Jupyter Notebooks or Google Colab, here is a useful guide to get started: `https://colab.research.google.com/notebooks/welcome.ipynb#scrollTo=GJBs_flRovLc`.

Summary

We have introduced location data and location data intelligence in this chapter by looking from different perspectives: business, technical, and data. We have also covered applications of location data intelligence and provided some simple and concrete examples from the Foursquare dataset. Here, both customer perspectives, as well as user business perspectives, were considered in our location data intelligence applications and examples. Furthermore, we have compared and contrasted data science and location data science. Finally, we have introduced a primer on using Jupyter Notebooks and Google Colab.

We will learn to process location data and apply machine learning models in the next chapter while consuming location data like a data scientist. We will use the New York taxi trajectory data to predict trip durations for New York taxicab trips.

2
Consuming Location Data Like a Data Scientist

Location comes in different forms, but what if it comes in a simple structured data format and we overlooked it all this time? Most machine learning algorithms, such as random forests, are geared toward creating insights from structured data in tabular form. In this chapter, we will discuss how to leverage spatial data that is masquerading as tabular data and apply machine learning techniques to it as any data scientist would. For this chapter, we will be using New York taxi trip data to predict trip duration for any given New York taxi trip. We are choosing this dataset because of the following reasons:

- Predicting trip duration has the right mix of geospatial analytics and machine learning
- Finding the time it takes to travel from point *A* to point *B* is a routing problem, which will be dealt with in `Chapter 6`, *Let's Build a Routing Engine*, and so this chapter is a perfect introduction

We will be using a library known as **fastai**, an amazing Python library built around popular machine learning libraries such as scikit-learn and PyTorch. In this chapter, we will be discussing the following topics:

- Exploratory data analysis
- Processing spatial data
- Understanding and inferring the error metric
- Building and inferencing a random forest model

Exploratory data analysis

For this chapter, we will be using curated data from the New York taxi trip dataset provided by the city of New York. The original source for this data can be found here: `https://data.cityofnewyork.us/api/odata/v4/hvrh-b6nb`.

 Visit the following website for more details about the data that's included in this dataset: `https://data.cityofnewyork.us/Transportation/2016-Green-Taxi-Trip-Data/hvrh-b6nb`.

For starters, let's have a peek at the data at hand using pandas. The curated data (`NYC_sample.csv`) that we will be using here can be found at the following download link: `https://drive.google.com/file/d/1OkkYZJEcsdCkU0V42eP6pj6YaK2WCGCE/view`.

```
df = pd.read_csv("NYC_sample.csv")
df.head().T
```

The curated New York taxi trip data that we are using has around 1.14 million records and has columns related to taxi fare, as well as trip duration, as you can see from the following screenshot:

	0	1	2	3	4
VendorID	2	2	2	2	1
lpep_pickup_datetime	05/07/2016 02:25:05 AM	05/07/2016 02:21:26 AM	05/08/2016 12:54:22 AM	05/08/2016 12:51:46 AM	05/07/2016 04:19:06 PM
Lpep_dropoff_datetime	05/07/2016 02:38:19 AM	05/07/2016 02:43:46 AM	05/08/2016 01:10:20 AM	05/08/2016 01:04:24 AM	05/07/2016 04:24:20 PM
Store_and_fwd_flag	N	N	N	N	N
RateCodeID	1	1	1	1	1
Pickup_longitude	-73.9589	-73.9234	-73.956	-73.9156	-73.9391
Pickup_latitude	40.7168	40.7069	40.8047	40.7433	40.8054
Dropoff_longitude	-73.9324	-74.0163	-73.9173	-73.9135	-73.9538
Dropoff_latitude	40.7079	40.7112	40.8244	40.7657	40.8063
Passenger_count	6	1	1	1	1
Trip_distance	2.25	6.34	3.3	2.21	1
Fare_amount	11	22	13.5	10.5	5.5
Extra	0.5	0.5	0.5	0.5	0
MTA_tax	0.5	0.5	0.5	0.5	0.5
Tip_amount	0	1	0	0	1.25
Tolls_amount	0	0	0	0	0
Ehail_fee	NaN	NaN	NaN	NaN	NaN
improvement_surcharge	0.3	0.3	0.3	0.3	0.3
Total_amount	12.3	24.3	14.8	11.8	7.55
Payment_type	1	1	2	1	1
Trip_type	1	1	1	1	1
PULocationID	NaN	NaN	NaN	NaN	NaN
DOLocationID	NaN	NaN	NaN	NaN	NaN

The data dictionary for this data that can be found at `https://data.cityofnewyork.us/api/views/hvrh-b6nb/files/65544d38-ab44-4187-a789-5701b114a754?download=true filename=data_dictionary_trip_records_green.pdf`. The data download page provides us with useful information about the data.

We can make the following inferences and assumptions about processing the data:

- Some columns have *missing values* (`NaN` included) that need to be handled.
- The value that we are trying to predict for each trip is the *trip duration*. This needs to be derived from the pickup and dropoff times for the training and validation data.
- Once the trip duration has been derived, we need to get rid of the *dropoff time* information since our objective is to compute the dropoff time when given other information, such as pickup location and time and dropoff location.
- We intend to primarily use the pickup and dropoff locations to predict time. Hence, in the training phase, we should be dropping records that don't have pickup and dropoff location information.
- We'd like to exclude columns related to *trip cost* because it doesn't contribute to the model.

We can do the following to tackle the problems we identified by taking a cursory glance at the data:

- Handle missing values
- Handle time values
- Handle unrelated data

Handling missing values

A machine learning algorithm such as random forest can handle a few missing values very well, and in some cases we can adopt strategies such as *imputing* or removing rows with missing values. But if the proportion of missing values in a column is pretty high, we might need to remove entire columns. The following lines of code help us determine the percentage of missing values in each column of the data:

```
na_counts = pd.DataFrame(df.isna().sum()/len(df))
na_counts.columns = ["null_row_pct"]
na_counts[na_counts.null_row_pct > 0].sort_values(by = "null_row_pct",
ascending=False)
```

The resulting `DataFrame` looks as follows:

	null_row_pct
Ehail_fee	1.000000
PULocationID	0.550597
DOLocationID	0.550597
Pickup_longitude	0.449403
Pickup_latitude	0.449403
Dropoff_longitude	0.449403
Dropoff_latitude	0.449403
Trip_type	0.000030

At first glance, we might be inclined to remove all rows that have missing latitude or longitude values for pickup and dropoff, since we identified that this is the major feature we will be building our model upon. But when taking closer look, we can see that the percentage of missing values for the `PULocationID` or `DOLocationID` columns and `Pickup_longitude`/`Pickup_latitude` and `Dropoff_longitude`/`Dropoff_latitude` are exact complements of each other. This means that the sum of the percentage values of entities; taking one from each group is exactly 100%. As a corollary, we can infer that for each missing value in pickup or dropoff coordinates, there is a non-missing value in the corresponding rows for `PULocationID` or `DOLocationID`.

But what are these location IDs? These location IDs are the taxi zone IDs that are assigned to different locations in New York. Though these locations are areal features, we can calculate the centroid of these locations and substitute these for the pickup and dropoff location coordinates. But when both the location ID and coordinates are missing, we need to remove those rows. The following lines of code will accomplish this:

```
df = df[~(
  (df.Dropoff_latitude.isna()) & (df.DOLocationID.isna())
)]
```

Handling time values

Time values play an important role in our model because time is both a feature and a target (value to be predicted) in our model. First, we need to convert the pickup and dropoff times into pandas `datetime` values to calculate the target value, which will be the natural log of the difference in time between dropoff and pickup in seconds:

```
df["trip_duration"] = np.log((df.Lpep_dropoff_datetime -
df.lpep_pickup_datetime).dt.seconds + 1)
```

In the preceding line of code, we are adding 1 second to the trip duration to prevent an undefined error when a log transformation is applied over the value.

But why are we using natural log transformation over the trip duration? There are three reasons for this, as follows:

- For the Kaggle competition on New York taxi trip duration prediction, the evaluation metric is defined as the **Root Mean Squared Logarithmic Error** (**RMSLE**). When log transformation is applied and the RMSE is calculated over the target values, we get the RMSLE. This helps us compare our results with the best-performing teams.
- Errors in log scale let us know by how many factors we were wrong, for example, whether we were 10% off from the actual values or 70% off. We will be discussing this in detail when we look at the *Error metric* section.
- The log transformation over the target variable follows a perfectly normal distribution. This satisfies one of the assumptions of linear regression. The plot of the trip duration values (on a log scale) looks as follows:

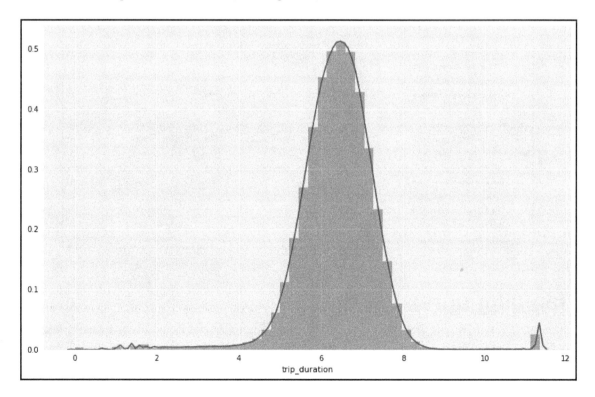

Time values as a feature

The pickup time (`pickup_datetime`) can be considered a feature, and can be further deconstructed into disparate components such as day of the week, month, hour, minute, and Boolean indicators, such as whether the `datetime` is a weekday or not, as shown in the following screenshot. `add_datepart()` is a convenience method in the `fastai.structured` module adds most of these components for us:

```
add_datepart(df, 'lpep_pickup_datetime', time=True)
df.tail().T
```

When we execute the preceding lines of codes, a host of `datetime` components are added to our `DataFrame`. The resulting `DataFrame` has a few extra columns, as we can see in the following snapshot of the `DataFrame`:

Payment_type	2	1	2	1	1
Trip_type	1	1	1	1	1
PULocationID	NaN	NaN	NaN	NaN	NaN
DOLocationID	NaN	NaN	NaN	NaN	NaN
lpep_pickup_datetimeYear	2016	2016	2016	2016	2016
lpep_pickup_datetimeMonth	5	5	5	5	5
lpep_pickup_datetimeWeek	18	18	18	18	18
lpep_pickup_datetimeDay	7	7	7	7	7
lpep_pickup_datetimeDayofweek	5	5	5	5	5
lpep_pickup_datetimeDayofyear	128	128	128	128	128
lpep_pickup_datetimeIs_month_end	False	False	False	False	False
lpep_pickup_datetimeIs_month_start	False	False	False	False	False
lpep_pickup_datetimeIs_quarter_end	False	False	False	False	False
lpep_pickup_datetimeIs_quarter_start	False	False	False	False	False
lpep_pickup_datetimeIs_year_end	False	False	False	False	False
lpep_pickup_datetimeIs_year_start	False	False	False	False	False
lpep_pickup_datetimeHour	16	16	16	2	2
lpep_pickup_datetimeMinute	24	20	20	2	24
lpep_pickup_datetimeSecond	5	59	50	17	26
lpep_pickup_datetimeElapsed	1462638245	1462638059	1462638050	1462586537	1462587866

Handling unrelated data

There are a few features that aren't related to the value to be predicted, in our case things such as fare and vendor ID and the dropoff time. Therefore, we will go ahead and drop these columns:

```
drop_columns = [
"Ehail_fee", "Extra", "Payment_type", "Total_amount",
```

```
"improvement_surcharge", "Tolls_amount", "Tip_amount", "MTA_tax",
"VendorID", "RateCodeID", "Store_and_fwd_flag", "Fare_amount",
"Lpep_dropoff_datetime", 'Trip_type ', 'Passenger_count'
]
df.drop(columns=drop_columns, inplace=True)
```

Spatial data processing

We will be discussing three things in this section: taxi zones, spatial joins, and the calculation of distances.

Taxi zones in New York

Analyzing and processing a taxi zone spatial data helps us achieve two objectives:

- Substitute the missing coordinates for pickup and dropoff locations with the taxi zone's centroid
- Use the taxi zone as a feature in the model

Visualization of taxi zones

We have provided the shapefile for the taxi zones in the data repository. Shapefiles can be read as (Geo)DataFrames with the Python library known as GeoPandas, like so:

```
taxi_zones = gpd.read_file("taxi_zones.shp")
taxi_zones.tail().T
```

We get the following output:

	258	259	260	261
OBJECTID	259	260	261	262
Shape_Leng	0.12675	0.133514	0.0271205	0.0490636
Shape_Area	0.000394552	0.000422345	3.43423e-05	0.00012233
zone	Woodlawn/Wakefield	Woodside	World Trade Center	Yorkville East
LocationID	259	260	261	262
borough	Bronx	Queens	Manhattan	Manhattan
geometry	POLYGON ((1025414.781960189 270986.1393638253,...	POLYGON ((1011466.966050446 216463.0052037984,...	POLYGON ((980555.2043112218 196138.486258477, ...	(POLYGON ((999804.7945504487 224498.5270484537...

The `GeoDataFrame` looks exactly like a pandas `DataFrame`, except that it has a special column named `geometry`. The `geometry` columns hold the area geometry of the taxi zones. We can visualize taxi zones with the plot method of the `GeoDataFrame`:

```
#Projecting Taxi Zones into WGS84 coordinate system
taxi_zones = taxi_zones.to_crs({'init': 'epsg:4326'})

#Plot the Geodataframe
ax = taxi_zones.plot(column = "zone", figsize = (12, 12), alpha = 0.4)
```

The following screenshot shows the visualization of taxi zones and a single trip:

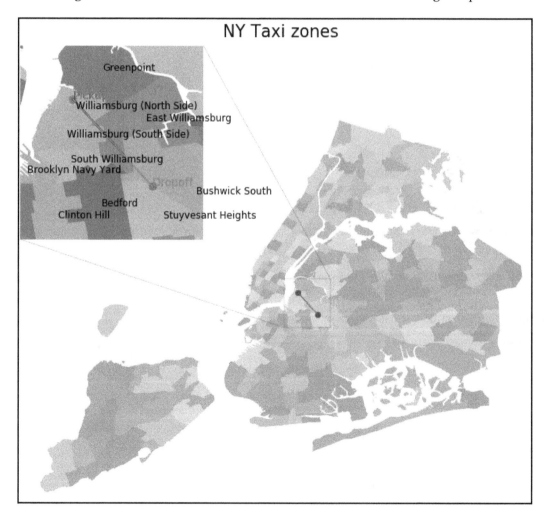

Spatial joins

If we can derive the pickup and dropoff taxi zone of each trip, we can add these as features to our machine learning model. For this, we need to perform an operation known as a spatial join, which is nothing but a Point-in-Polygon solution that's supported by the GeoDataFrame. The following code has an `assign_taxi_zones()` function, which takes our DataFrame as an input and returns a pandas series. Internally, it does three things:

- Construct a GeoDataFrame using the input DataFrame's latitude and longitude values (point geometry)
- Perform a spatial join between the point and the taxi zones (polygon geometry)
- Return the location ID of the taxi zone for each coordinate

These steps are implemented in the following lines of code:

```python
from shapely.geometry import Point

def assign_taxi_zones(df, lon_var, lat_var, locid_var):
    try:
        # Construct a Geodataframe using the coordinates of each trip
        local_gdf = gpd.GeoDataFrame(
            crs={'init': 'epsg:4326'},
            geometry=[Point(xy) for xy in
                      zip(df[lon_var], df[lat_var])])

        #Perform a spatial join with the Taxi Zones
        local_gdf = gpd.sjoin(local_gdf, taxi_zones, how='left',
op='within')
        return local_gdf.LocationID.rename(locid_var)
    except ValueError as ve:
        print(ve)
        print(ve.stacktrace())
        series = df[lon_var]
        series = np.nan
        return series

"""
Calculate pickup and dropoff taxi zone ids
"""

df['pickup_taxizone_id'] = assign_taxi_zones(df,
"Pickup_longitude","Pickup_latitude", "pickup_taxizone_id")

df['dropoff_taxizone_id'] = assign_taxi_zones(df,
"Dropoff_longitude","Dropoff_latitude","dropoff_taxizone_id")
```

Remember, this operation will only assign a taxi zone that we know the coordinates of.

To backfill the missing coordinates with the centroid of the taxi zone, we can follow these steps:

1. Find the centroid of each taxi zone
2. Join the `DataFrame` with the taxi zones based on the pickup zone ID that we just computed (`pickup_taxizone_id`)
3. Transfer the taxi zone's centroid to the `DataFrame`
4. For all rows with missing pickup coordinates, substitute the centroid values
5. Apply the same process in order to backfill missing dropoff coordinates

The following lines of code illustrate this process:

```
#1. Finding Taxi Zone' Centroid
taxi_zones["X"] = taxi_zones.centroid.x
taxi_zones["Y"] = taxi_zones.centroid.y

#2. Join dataframe with taxizone based on pickup zone id
df = pd.merge(df, taxi_zones[["LocationID","X", "Y"]], how = "left",
left_on = "PULocationID", right_on = "LocationID")

#3.Substitute missing lat/long values w/
# the taxi zone's centroid

df.Pickup_longitude.fillna(df.X, inplace=True)
df.Pickup_latitude.fillna(df.Y, inplace=True)


df.drop(columns=["LocationID", "X", "Y"], inplace=True)

#5. Apply same process for Dropoff zone
df = pd.merge(df, taxi_zones[["LocationID","X", "Y"]], how = "left",
left_on = "DOLocationID", right_on = "LocationID")

df.Dropoff_longitude.fillna(df.X, inplace=True)
df.Dropoff_latitude.fillna(df.Y, inplace=True)

df.drop(columns=["LocationID", "X", "Y"], inplace=True)

df.tail().T
```

We can use a very similar process to add the borough names from the taxi zone shapefile to each row in the DataFrame. In this process, we have added the following features:

- Pickup zone ID
- Dropoff zone ID
- Pickup borough
- Dropoff borough

We were also quite successful in backfilling many missing values in pickup and dropoff location coordinates, as well as taxi zone IDs.

Calculating distances

When it comes to distance, there are different kinds of distance that make sense in this context. These are as follows:

- Distance as the crow flies (or Haversine distance)
- Distance as you drive in Manhattan (or Manhattan distance)

Haversine distance

Haversine distance is the **Great Circle Distance** (**GCD**) between two geographic coordinates. A GCD is incidentally the shortest distance between the two coordinates. This is almost similar to a Euclidean distance (or a straight-line distance), except that we are accounting for the spherical nature of the Earth (yes, we are generalizing the Earth as a sphere with a radius of 3,958 miles to make our lives easier). The Python code for calculating the Haversine distance is as follows:

```
import numpy as np

def haversine(lat1, lon1, lat2, lon2):
    R = 3958.76 # Earth radius in miles
    dLat = np.radians(lat2 - lat1)
    dLon = np.radians(lon2 - lon1)
    lat1 = np.radians(lat1)
    lat2 = np.radians(lat2)
    a = np.sin(dLat/2) ** 2 + np.cos(lat1) * np.cos(lat2) * np.sin(dLon/2) ** 2
    c = 2*np.arcsin(np.sqrt(a))
    return R * c
```

The preceding function takes a pair of coordinates (the latitude and longitude values of the source and destination, respectively) and returns the Haversine distance between them.

Manhattan distance

Manhattan distance is a distance metric inspired by the near rectangular street block design in Manhattan. The distance between two locations is calculated as the sum of the straight-line distance along the x axis and the straight-line distance along the y axis:

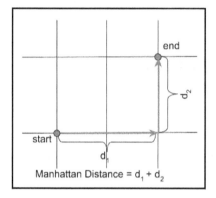

Here, the x axis distance (**d1**) is nothing but the difference in longitude, while the y axis distance (**d2**) is the Haversine distance of the difference in latitude. Before we proceed with our calculation, there's just one small geographical subtlety that we have to take care of, and that is the angle made by the north-south running streets of New York to the true north. This angle was established as around 28.9° clockwise from true north. For more details on the derivation of this angle, please refer to this excellent work by Charles Petzold: http://www.charlespetzold.com/etc/AvenuesOfManhattan/:

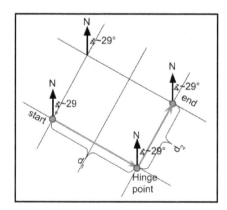

In this case, to calculate the Manhattan distance, we need to find the coordinates of the hinge point in a coordinate system and transform them by a clockwise rotation of 29°. If the north-south running streets aligned with true north, the hinge coordinates would be as follows:

Longitude of hinge point = longitude of dropoff point

Latitude of hinge point = latitude of the pickup point

Manhattan Distance = Haversine distance between pickup and hinge point + Haversine distance between the hinge and dropoff point

The transformation of coordinates that's achieved by rotating a coordinate axis is given by a rotation matrix:

$$R(\theta) = \begin{bmatrix} cos\theta & -sin\theta \\ sin\theta & cos\theta \end{bmatrix}$$

The following operations are required to calculate the Manhattan distance when given pickup and dropoff coordinates:

1. Perform a matrix multiplication of the pickup and dropoff coordinates with the rotation matrix, where θ = -29º
2. Derive the hinge point coordinates as per the formula provided
3. Transform the hinge point back into the geographic coordinate system by performing a rotation of an equal amount in an anti-clockwise direction (θ = +29º)
4. Use the preceding formula to calculate the Manhattan distance using the pickup, hinge, and dropoff coordinates

The Python code which accomplishes this is as follows:

```
theta1 = np.radians(-28.904)
theta2 = np.radians(28.904)
R1 = np.array([[np.cos(theta1), np.sin(theta1)], [-np.sin(theta1),
np.cos(theta1)]])
R2 = np.array([[np.cos(theta2), np.sin(theta2)], [-np.sin(theta2),
np.cos(theta2)]])

def manhattan_dist(lat1, lon1, lat2, lon2):
    p = np.stack([lat1, lon1], axis = 1)
    d = np.stack([lat2, lon2], axis = 1)
```

```
    pT = R1 @ p.T
    dT = R1 @ d.T

    vT = np.stack((pT[0,:], dT[1,:]))
    v = R2 @ vT
    return (haversine(p.T[0], p.T[1], v[0], v[1]) + haversine(v[0], v[1],
d.T[0], d.T[1]))
```

For now, let's just add the first two types of distance as different columns to the training `DataFrame`:

```
df["haversine_dist"] = haversine(df["pickup_latitude"],
df["pickup_longitude"], \
 df["dropoff_latitude"], df["dropoff_longitude"])

df["manhattan_dist"] = manhattan_dist(df["pickup_latitude"],
df["pickup_longitude"], \
 df["dropoff_latitude"], df["dropoff_longitude"])
```

Error metric

If we visit the evaluation section of the Kaggle competition, the evaluation metric is defined as the RMSLE. In the competition, the objective is to minimize this metric for the test data. An error is simply the difference between actual values and predicted values:

$$error = predicted\ value - actual\ value$$

The **Root Mean Squared Error** (**RMSE**) would literally be the square root applied over the mean of all the squared error terms for each observation.

However, our metric in the Kaggle competition needs to be a log error:

$$log\_error = log(predicted\ value + 1) - log(actual\ value + 1)$$

Therefore, it is important to apply a log transform over the `trip_duration` column as we did earlier:

```
df["trip_duration"] = np.log(df["trip_duration"] + 1)
```

Now, we can use a function that can calculate RMSE rather a function that calculates RMSLE:

```
import math
def rmse(x,y): return math.sqrt(((x-y)**2).mean())
```

Interpreting errors

What does an RMSLE of, say, 0.3 actually mean? Well, let's visit the formula for the log error again:

$$log\_error = 0.3$$

$$log(predicted\_value + 1) - log(actual\_value + 1) = 0.3$$

$$e^{log(\textbf{predicted\_value}+1/\textbf{actual\_value}+1)}=e^{0.3}$$

$$log((predicted\_value + 1) / (actual\_value + 1)) = 0.3$$

$$(predicted\_value + 1) / (actual\_value + 1) = 1.349$$

$$predicted\_value = 1.349 * actual\_value + 0.349$$

If the preceding derivation is hard to follow, it doesn't matter; it's just for math enthusiasts. What this means is that, on average, our naive model predicts the trip duration, 1.35 times the actual value. This is not too bad, given that the best model in the Kaggle competition predicts, on average, 1.3 times the actual value. We can arrive at this metric by using a single line of Python code:

```
np.exp(rmsle)
```

The response to the preceding line of code is the factor by which our predictive model is off from the actual values.

Building the model

Let's build the final model using a random forest regressor. A random forest is a universal machine learning technique, that is, it can handle different kinds of data; it could be a category (classification), a continuous variable (regression), or features of any kind, such an image, price, time, post codes, and so on (that is, both structured and unstructured data). It doesn't generally overfit too much, and it is very easy to stop it from overfitting. For these reasons, random forest is a versatile ML technique which we can effectively use to solve our problem.

Validation data and error metrics

Our initial step is choosing a suitable size for validation data. Before delineating the validation dataset and defining accuracy metrics, we have just two more steps to take into account that will make our data ready for building models. These are two convenience functions that are provided by fastai to make our models more robust:

- `train_cats()`: Convert any string data into categorical data
- `proc_df()`: Perform one-hot encoding on categorical variables and handle missing values

Let's have a look at the following code snippet:

```
train_cats(train_df)
tdf, y, nas = proc_df(df, 'trip_duration')
```

A validation data size of 20,000 will be enough to validate our model:

```
def split_vals(a,n):
    return a[:n].copy(), a[n:].copy()

n_valid = 20000
n_trn = len(tdf)-n_valid
raw_train, raw_valid = split_vals(df, n_trn)
X_train, X_valid = split_vals(tdf, n_trn)
y_train, y_valid = split_vals(y, n_trn)

X_train.shape, y_train.shape, X_valid.shape
```

We will use RMSE and R2 as accuracy metrics. RMSE is a very simple yet effective measure to understand errors, and R2 is a very effective metric to evaluate the predictive power of a model. In the Kaggle competition, we talked about the top RMSE values, which are around 0.28 at the time of writing:

```
def rmse(x,y): return np.sqrt(((x-y)**2).mean())

def print_score(m):
    res = f"""Train RMS : {rmse(m.predict(X_train), y_train)},
            Valid RMSE : {rmse(m.predict(X_valid), y_valid)},
            Train R2 score : {m.score(X_train, y_train)},
            Valid R2 score: {m.score(X_valid, y_valid)}
            """
    if hasattr(m, 'oob_score_'): res += f" OOB Score : {m.oob_score_}"
    print(res)
```

Let's go ahead and run our first model with about 40 estimators (trees). We will be using all the CPUs that are available to us to enable multiprocessing in the background (hence the n_jobs = -1 parameter):

```
m = RandomForestRegressor(n_estimators = 40, n_jobs=-1, oob_score=True)
m.fit(X_train, y_train)
print_score(m)
```

The preceding model gives us these scores. A validation RMSE of 0.25 indicates that we are among the 5th percentile of competitors:

```
Train RMS : 0.09880475531046715,
Valid RMSE : 0.2599229455143868,
Train R2 score : 0.9786976405479708,
Valid R2 score: 0.8602128925971101
OOB Score : 0.8480746084914459
```

Let's see if we can make the model any better. There are some cool methods in fastai that let us understand the importance of the features that are used in the model, as well as the correlation among the features (multicollinearity). For example, we can easily look at the important features in our model by using the rf_feature_importance() method:

```
fi = rf_feat_importance(m, df_trn); fi[:20]feature_imp
def plot_fi(fi): return fi.plot('cols', 'imp', 'barh', figsize=(16,8),
legend=False, grid = False)

plot_fi(fi[:20]);
```

The preceding code also provides a nice snippet so that we can display the feature's importance as a bar chart, as follows:

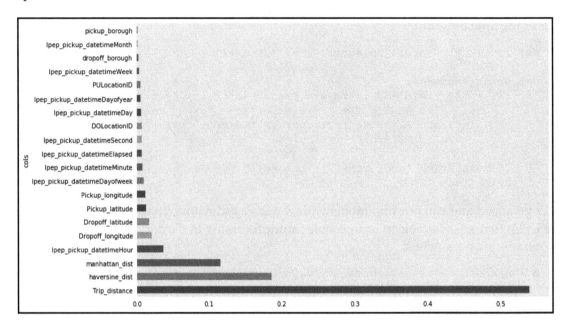

We can also write our own code to assess conditions such as multicollinearity, in which two features are closely associated with one another. The following lines of code plot a dendrogram, which shows multicollinearity between features:

```
from scipy.cluster import hierarchy as hc

corr = np.round(scipy.stats.spearmanr(df_keep).correlation, 4)
corr_condensed = hc.distance.squareform(1-corr)
z = hc.linkage(corr_condensed, method='average')
fig = plt.figure(figsize=(12,10))
dendrogram = hc.dendrogram(z, labels=df_keep.columns, orientation='left',
leaf_font_size=12)
plt.show()
```

The dendrogram shows highly correlated features, such as `Dayofyear` and `datetimeElapsed`, as well as `datetimeMonth`. The analysis also shows high correlation between the Manhattan and Haversine distances:

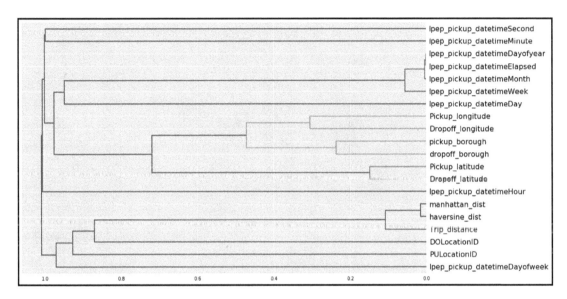

Once we remove such redundant features and introduce the regularization parameter (`max_features = 0.5`), our model's RMSE drops even further, so it's among the top 2%ile of the leaderboard!

Summary

In this chapter, we chose a pertinent problem that had both analytics and geospatial components and tried to apply a very robust ML technique known as random forest to it. Before building the model, we had to handle the date component, the spatial component of data, as well as the categorical and continuous variables. We were able to achieve a good score in our first pass and build a world-class model with a few lines of code and a little bit of spatial data processing.

In the next chapter, we will discuss more accurate real-world distance metrics and perform other spatial computations, such as intersection, to make the model better.

3
Performing Spatial Operations Like a Pro

Spatial operations play a crucial role in location data analysis. In fact, they are what distinguishes location data analysis from other data analysis. Spatial analysis derives insights from input data using geometry functions and operators. There are a number of different spatial operations, including buffer analysis, spatial joins, overlay analysis, and topological operations (such as intersect and contain). We will use the Foursquare dataset to apply spatial operations in a real-world project.

 The NYC Foursquare Check-In dataset first appeared in *Fine-Grained Preference-Aware Location Search Leveraging Crowdsourced Digital Footprints* by *Dingqi Yang, Daqing Zhang, Zhiyong Yu*, and *Zhiwen Yu*, from *LBSNs*. This was released before the 2013 *ACM International Joint Conference on Pervasive and Ubiquitous Computing* (*UbiComp 2013*), September 8 to 12, 2013, in Zurich, Switzerland.

As we have briefly discussed and given an overview of the importance of the NYC Foursquare check-in dataset in Chapter 1, *Introducing Location Intelligence*, we will continue to derive insights using the GeoPandas library, which you were briefly introduced to in Chapter 2, *Consuming Location Data Like a Data Scientist*. We will dive deep and use GeoPandas to create **geographic DataFrames** (**GeoDataFrames**) from pandas data, extract geometries, apply geographic coordinate projections, and perform spatial operations. At the end of the chapter, we will study interactive location data visualization with Folium.

In this chapter, we will learn about the following topics:

- Consuming location data as GeoDataFrames
- Extracting geometries from spatial data
- Performing spatial operations on GeoDataFrames
- Understanding and applying topological operators
- Interactive location data visualization

GeoDataFrames and geometries

In this section, we will learn the basics of loading and processing location data using the GeoPandas library. GeoPandas is built on top of the pandas and NumPy libraries. Like pandas, GeoPandas' data structure contains GeoDataFrames and GeoSeries. GeoPandas provides not only the capability to read and manipulate geographic data easily, but it can also perform many essential geospatial operations, including geometric and spatial operations, topological analysis, and geographic projections. You can also visualize and plot maps with GeoPandas (providing a high-level interface to the matplotlib library) by using the `plot()` method on a GeoDataFrame or GeoSeries. An important technique in location data analysis is the ability to convert simple CSVs or pandas DataFrames to GeoDataFrames, and that can be easily done in GeoPandas.

Before we delve deep into GeoDataFrames and how to convert CSVs into GeoDataFrames, we need to learn about what geometry is and how important it is to location data.

Geographic coordinates and geometries

Geographic coordinates and geometry objects form the glue that holds together all location data. Geometry objects define shapes in spatial locations: points, lines, polygons, and their constituent parts. Put another way, we can say that geographic coordinates tell us about the locations while geometries hold their shapes. In any spatial operations, we need to have both geometry and coordinates to carry out a meaningful geographic analysis.

Let's look at an example from the Foursquare data. We will use Google Colab Jupyter Notebook to perform spatial operations in this chapter. The dataset for this chapter is available in the `Chapter03` folder of the notebook and can also be accessed directly from Dropbox at: `https://www.dropbox.com/s/0zytrf2ncoquxgq/Foursquare_2014_NYC.zip`. You can go ahead and upload the data as instructed in `Chapter 1`, *Introducing Location Intelligence*; however, I will also show you how to access it directly in Google Colab with the `wget` functionality. It will be useful to you to learn how to access data on the web, so we will use `wget` in this section. Really, `wget` is a great utility for accessing files from the web, and supports a number of different protocols. To get the data into the Google Colab environment, we simply pass the URL and `wget`.

Accessing the data

To access the data directly with `wget`, we just need to provide the URL of the data. In this case, we can directly provide the link to the dataset that was provided earlier:

```
!wget https://www.dropbox.com/s/0zytrf2ncoquxgq/Foursquare_2014_NYC.zip
```

Now that we have our data inside the Google Colab environment, we can unzip it:

```
# let us unzip it
!unzip Foursquare_2014_NYC.zip
```

We will first use pandas to read the CSV file and later convert it to a GeoPandas `GeoDataFrame`:

```
col_names = ['UserID', 'VenueID', 'VenueCategoryID',
             'VenueCategoryName', 'Latitude', 'Longtitude',
             'Timezone', 'UTCtime']
nyc =
pd.read_csv('Foursquare/dataset_tsmc2014/dataset_TSMC2014_NYC.txt',names=co
l_names,sep="\t", encoding = "ISO-8859-1" )
nyc.head()
```

The following are the first five rows of the pandas `DataFrame`:

	UserID	VenueID	VenueCategoryID	VenueCategoryName	Latitude	Longtitude	Timezone	UTCtime
0	470	49bbd6c0f964a520f4531fe3	4bf58dd8d48988d127951735	Arts & Crafts Store	40.719810	-74.002581	-240	Tue Apr 03 18:00:09 +0000 2012
1	979	4a43c0aef964a520c6a61fe3	4bf58dd8d48988d1df941735	Bridge	40.606800	-74.044170	-240	Tue Apr 03 18:00:25 +0000 2012
2	69	4c5cc7b485a1e21e00d35711	4bf58dd8d48988d103941735	Home (private)	40.716162	-73.883070	-240	Tue Apr 03 18:02:24 +0000 2012
3	395	4bc7086715a7ef3bef9878da	4bf58dd8d48988d104941735	Medical Center	40.745164	-73.982519	-240	Tue Apr 03 18:02:41 +0000 2012
4	87	4cf2c5321d18a143951b5cec	4bf58dd8d48988d1cb941735	Food Truck	40.740104	-73.989658	-240	Tue Apr 03 18:03:00 +0000 2012

First five rows of the Foursquare dataset

We have both nonspatial attributes, such as `UserID` and `VenueID`, and also spatial attributes, such as `Latitude` and `Longitude`. In fact, the latitude and longitude attributes represent our geometries. Each point is usually represented as a pair of coordinates (`Latitude`, `Longitude`), and by being combined, they form a point geometry. Let's see how we can create a geometry out of the `Latitude` and `Longitude` columns. We will loop through both columns using the `zip` function in Python and simply put them in tuple format.

Geometry

The geometry can be created from latitude and longitude. We need to use the Shapely library when creating geometries, but for now, let's loop the first five rows and create a tuple out of them, as follows:

```
for lat, lon in zip(nyc['Latitude'][:5], nyc['Longtitude'][:5]):
    geometry = lat, lon
    print(geometry)
```

The output will be as follows:

```
(40.71981037548853, -74.00258103213994)
(40.606799581406435, -74.04416981025437)
(40.71616168484322, -73.88307005845945)
(40.7451638, -73.982518775)
(40.74010382743943, -73.98965835571289)
```

This is a simple `Latitude` and `Longitude` tuple, but we want to create a spatial `geometry` column that is aware of both the shape and coordinates. As such, we will use the Shapely library, which provides a lot of spatial geometry operations and uses many of GeoPandas' geometric operations under the hood. Now we will do the same as before, but this time, we will import the Shapely `geometry` point and wrap around the `Latitude` and `Longitude` tuple, as follows:

```
from shapely.geometry import Point
for lat, lon in zip(nyc['Latitude'][:5], nyc['Longtitude'][:5]):
    geometry = Point(lat, lon)
    print(geometry)
```

The output will be as follows:

```
POINT (40.71901037548853 -74.00258103213994)
POINT (40.60679958140643 -74.04416981025437)
POINT (40.71616168484322 -73.88307005845945)
POINT (40.7451638 -73.982518775)
POINT (40.74010382743943 -73.98965835571289)
```

By wrapping around the Shapely point geometry, we create a `geometry` point. In Shapely, we can, in fact, create different geometry shapes, including lines and polygons, and thus enable our data to perform spatial operations. For example, we can simply create a buffer around each point to mark it out, get its coordinates, and plot it on a map. But before we do that, we need to specify a coordinate system for these points.

Coordinate reference systems

Geographic **coordinate reference system** (**CRS**) enable us to define and specify every location on Earth by using three dimensional spherical surfaces. Geographic CRS and projection from one reference to another are crucial in many location data analysis applications. Here, we will cover the foundational skills that you need to master for this. Normally, we use a code number to indicate the CRS of geographic data. Standardized CRS codes are available from the **European Petroleum Survey Group** (**EPSG**). The most commonly used EPSG codes include `epsg:4326` and `epsg:3395` for world data projects. Usually, each country or region has a different local CRS that you can search by country name or EPSG from `http://epsg.io/` or `http://spatialreference.org`. We will perform projections later, but for now, we will simply use `epsg:4326` to demonstrate how we can create a `GeoDataFrame`, as follows:

```
crs = {'init': 'epsg:4326'}
```

GeoDataFrames

Now we can easily create a `GeoDatFrame` since we have both geometries and a CRS. To create a `GeoDataFrame`, you need to pass a `DataFrame`, `geometry`, and optionally a CRS:

```
nyc_gdf = gpd.GeoDataFrame(nyc, crs=crs, geometry=geometry)
nyc_gdf.head()
```

Once we convert the `DataFrame` into a `GeoDataFrame`, we will have an additional column called `geometry`, as you can see from the following screenshot:

	UserID	VenueID	VenueCategoryID	VenueCategoryName	Latitude	Longtitude	Timezone	UTCtime	geometry
0	470	49bbd6c0f964a520f4531fe3	4bf58dd8d48988d127951735	Arts & Crafts Store	40.719810	-74.002581	-240	Tue Apr 03 18:00:09 +0000 2012	POINT (40.71981037548853 -74.00258103213994)
1	979	4a43c0aef964a520c6a61fe3	4bf58dd8d48988d1df941735	Bridge	40.606800	-74.044170	-240	Tue Apr 03 18:00:25 +0000 2012	POINT (40.60679958140643 -74.04416981025437)
2	69	4c5cc7b485a1e00d35711	4bf58dd8d48988d103941735	Home (private)	40.716162	-73.883070	-240	Tue Apr 03 18:02:24 +0000 2012	POINT (40.71616168484322 -73.88307005845945)
3	395	4bc7086715a7ef3bef9878da	4bf58dd8d48988d104941735	Medical Center	40.745164	-73.982519	-240	Tue Apr 03 18:02:41 +0000 2012	POINT (40.7451638 -73.982518775)
4	87	4cf2c5321d18a143951b5cec	4bf58dd8d48988d1cb941735	Food Truck	40.740104	-73.989658	-240	Tue Apr 03 18:03:00 +0000 2012	POINT (40.74010382743943 -73.98965835571289)

The Foursquare dataset converted into a GeoDataFrame (geometry column)

GeoPandas automatically adds a `geometry` column for our data, which enables a lot of spatial operations, as we will see in the following sections and throughout this book. We can also use this `geometry` column to easily plot our data as a map. In GeoPandas, this can be achieved with the `.plot()` function.

We first prepare the plot by creating a figure and axes to plot the map:

```
fig, ax = plt.subplots(figsize=(12,10))
```

Plot `nyc_gdf` `GeoDataFrame` as a map. Here, we pass `markersize` as `0.1` and provide the axis we just created:

```
nyc_gdf.plot(markersize=0.1,ax=ax);
```

The following code removes both the x and y tick labels, sets the title of the map as Foursquare Points, and finally tells Matplotlib to show the map inline in the notebook:

```
ax.set_xticklabels([])
ax.set_yticklabels([])
ax.set_title('Foursquare Points')
plt.show()
```

Here is the map of all the points in our nyc_gdf GeoDataFrame:

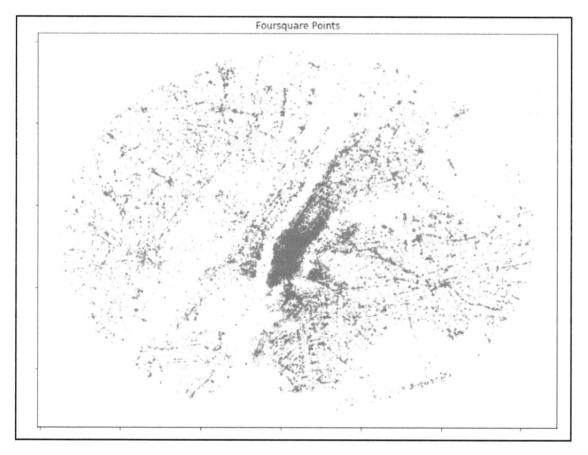

NYC Foursquare points mapped

Now that we have a GeoDataFrame, we can perform spatial operations, including projections, buffer analysis, and spatial joins, which we will cover in the next section.

Spatial operations

In this section, we will use the `GeoDataFrame` point that we just created and the polygons of the NYC census tracts to demonstrate different spatial operations. Before we can do any meaningful location data analysis, we need to check the geographic CRS of our data. As we have already mentioned, we have WGS84, and the coordinates are defined as decimal degrees.

Projections

It is a common process to reproject data from one format, such as WGS84, to other formats. There are many different projections; some distort shapes, others distort size, while other projections maintain an equal area size. Doing so is very useful for visualizing how different projections transform data, as is made clear at `https://map-projections.net/imglist.php`. In our case, we look for a metric projection suitable for NYC. The official EPSG website (`http://epsg.io/`) is very helpful when searching for relevant projections of your data via coordinates, location name, ZIP codes, or EPSG code, if you know it. We will choose `epsg:32618` projection (unit of measurement in this projection is meter) that covers the NYC zone. Reprojecting our data is easy with GeoPandas, and we can simply pass the EPSG code to the `to_crs` function:

```
nyc_gdf_proj = nyc_gdf.to_crs({'init': 'epsg:32618'})
nyc_gdf_proj.head()
```

The output will be as follows:

	UserID	VenueID	VenueCategoryID	VenueCategoryName	Latitude	Longtitude	Timezone	UTCtime	geometry
0	470	49bbd6c0f964a520f4531fe3	4bf58dd8d48988d127951735	Arts & Crafts Store	40.719810	-74.002581	-240	Tue Apr 03 18:00:09 +0000 2012	POINT (584239.3260128839 4508132.469739441)
1	979	4a43c0aef964a520c6a61fe3	4bf58dd8d48988d1df941735	Bridge	40.606800	-74.044170	-240	Tue Apr 03 18:00:25 +0000 2012	POINT (580863.1796990195 4495548.605023375)
2	69	4c5cc7b485a1e21e00d35711	4bf58dd8d48988d103941735	Home (private)	40.716162	-73.883070	-240	Tue Apr 03 18:02:24 +0000 2012	POINT (594338.2357704557 4507848.94936957)
3	395	4bc7086715a7ef3bef9878da	4bf58dd8d48988d104941735	Medical Center	40.745164	-73.982519	-240	Tue Apr 03 18:02:41 +0000 2012	POINT (585901.1405895475 4510966.314499039)
4	87	4cf2c5321d18a143951b5cec	4bf58dd8d48988d1cb941735	Food Truck	40.740104	-73.989658	-240	Tue Apr 03 18:03:00 +0000 2012	POINT (585304.8352019053 4510397.658739946)

Projected GeoDataFrame

If you look closely and compare the `geometry` columns of the projected data (`nyc_gdf_proj`) and nonprojected data (`nyc_gdf`), you will realize that they are in different formats. Because of the projection, our data has been transformed from decimal degrees into a meter-based format.

Buffer analysis

Now that we are able to analyze our location data and get insights based on meter units, let's explore buffer analysis. Buffer analysis is one of the most used GIS spatial operations. It creates zones with a certain area around a point, line, or polygon geometry according to a specified buffer distance. For example, if we take one point from our projected `GeoDataFrame`, we can create buffers of 10, 50, and 100 meters around the first point in our data:

```
point1 = nyc_gdf_proj[:1]
buf10 = point1.buffer(10)
buf50 = point1.buffer(50)
buf100 = point1.buffer(100)
```

Now, let's plot all the buffered points and the original point together. We pass all images into the same axis to overlay on top of each other:

```
fig, ax = plt.subplots(figsize=(12, 10))
buf100.plot(color = 'red', ax=ax);
buf50.plot(ax=ax, color='yellow')
buf10.plot(ax=ax, color='gray');
point1.plot(ax=ax, color='black')
ax.set_xticklabels([])
ax.set_yticklabels([])
plt.show()
```

As you can see from the following map, the small dot inside the circles is the original point, `point1`. The small circle in the first tier around the original point represents the 10-meter buffer, while the next tier, the middle circle, represents the 50-meter buffer. The largest outer circle is the 100-meter buffer:

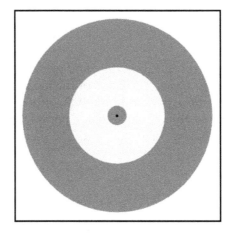

Buffered point: 10, 50, and 100 meters

Let's explore further how we can use buffer analysis for a subset of our data. We might be interested in seeing only subway data and perform a buffer analysis for that. We will take only data where `['VenueCategoryName']== 'Subway'`, provide a distance (`1000` meters) by using `.buffer`, and then plot it.

For example, say we are interested in how close subway stations are to each other. Although we can calculate distance as we did before, we can also try buffer analysis to see whether restaurants are near to each one or overlap in their service areas. In this example, we first take a subset of our data, the subway points, and then a buffer 1,000 meters around each point:

```
subway = nyc_gdf_proj[nyc_gdf_proj['VenueCategoryName']== 'Subway']
subwayBuf = subway.buffer(1000)
```

We then plot the `subwayBuf` polygons with points:

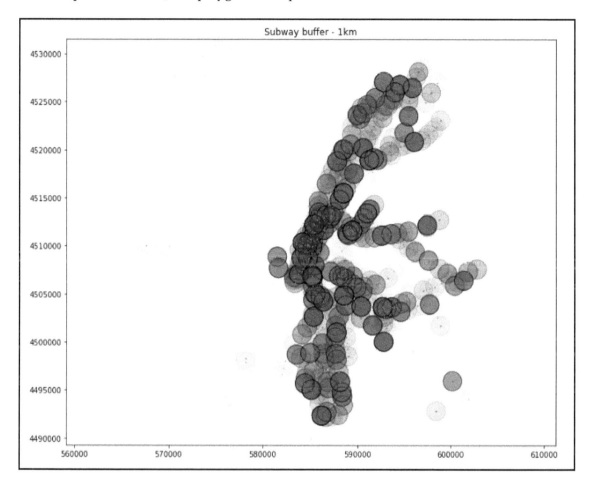

Subways buffered by 1,000 meters (shown in blue); the original points are the red dots

Spatial joins

The table join is a classical query operation where two separate tables sharing a column (foreign ID) are merged based on that column. The table join does not involve any geographic relations, but only involves table attributes; however, we can use GeoDataFrame options to perform a spatial join, merging two geometry objects based on their locations. Let's look at an example of this. We will add a new dataset of NYC districts with polygon geometry. We will access the data directly from the server URL as GeoJSON and look at the first five rows, as follows:

```
url_dist =
'http://services5.arcgis.com/GfwWNkhOj9bNBqoJ/arcgis/rest/services/nyad/Fea
tureServer/0/query?where=1=1&outFields=*&outSR=4326&f=geojson'
nyc_dist = gpd.read_file(url_dist)
nyc_dist.head()
```

The output will be as follows:

	OBJECTID	AssemDist	Shape__Area	Shape__Length	geometry
0	1	34	5.549695e+07	60366.244237	POLYGON ((-73.89191238093809 40.7657193793578,...
1	2	47	6.717310e+07	47467.275604	POLYGON ((-73.97299577171449 40.6088223789542,...
2	3	48	7.374100e+07	51820.347851	POLYGON ((-73.9609192013542 40.627682756256, -...
3	4	49	6.114401e+07	54424.138562	POLYGON ((-73.9998906540922 40.6379042694496, ...
4	5	54	7.937047e+07	73969.610179	POLYGON ((-73.90877750768648 40.6982634836477, ...

NYC districts: the first five rows

If we run `nyc_dist.plot`, the map for the district data looks like the following.

You might realize from the `geometry` column or the *x* and *y* axes of the following map that we are using decimal degrees of latitude and longitude, and if we check the CRS, we get `epsg:4326`, which we need to convert to a meter-based projection, just like the NYC Foursquare data:

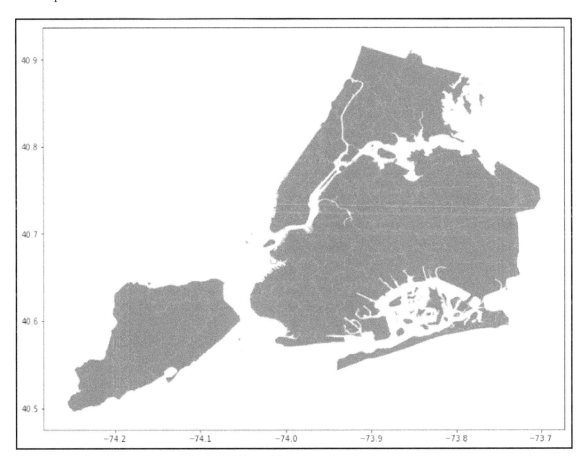

NYC districts polygon map

To convert a `GeoDataFrame` CRS, GeoPandas has the `to_crs` function, which takes a dictionary of the projection to be used. We will use `epsg:32618` here, as shown in the following code:

```
# Convert to UTM meter based projection: https://epsg.io/32618
nyc_dist_proj = nyc_dist.to_crs({'init': 'epsg:32618'})
```

Now that both datasets have the same CRS, let's overlay the points data and the polygon data to check whether their locations match. The locations match, but as you can see, the Foursquare dataset is actually not only in NYC, but also extends beyond NYC district boundaries. It is not uncommon to have such scenarios where you have different data that does not perfectly fit with your desired boundaries.

The following code overlays both the points and boundary datasets. Once we create the figure and axis, we can easily pass any plot to the same axis to overlay it:

```
fig, ax = plt.subplots(figsize=(12,12))
nyc_dist_proj.plot(ax=ax, color='gray');
nyc_gdf_proj.plot(ax=ax, markersize=0.01, color='black');
```

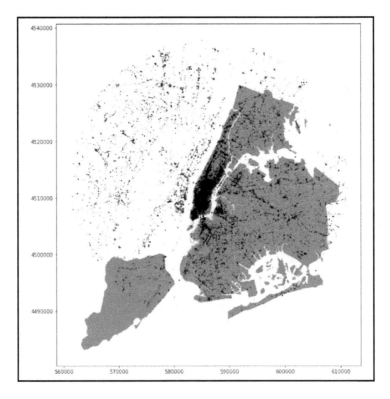

NYC districts projected and Foursquare points overlaid

Here is where the spatial join helps in your location data preprocessing and analysis. We will first choose points that fall within NYC district boundaries based on their locations using spatial join operations. A spatial join is when two geometry objects are merged based on their spatial relationship. In GeoPandas, we can carry out a spatial join with the `.sjoin` method, which takes two GeoDataFrames and an operation type. The operation type determines the type of join to apply. It could be an `intersection`, `within`, or `contains` operation, and can be carried out with different geometries, such as points with polygons, lines with polygons, or points with lines.

To illustrate this, let's get points that are only within NYC districts. Here is how this would be performed in code: we use `.sjoin` and pass the two GeoDataFrames, `nyc_gdf_proj` and `nyc_dist_proj`. We also need to provide the operation; here, the `within` operation means we get only those points *within* the boundaries:

```
nyc_points = gpd.sjoin(nyc_gdf_proj, nyc_dist_proj, op='within')
```

Now, if we look at the overlay map with `nyc_points` only, we can see that we have excluded all other points and only have points within NYC boundaries:

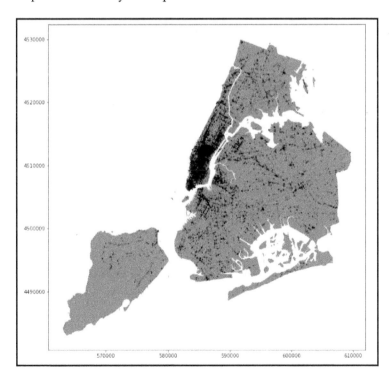

Spatial join: NYC districts projected and Foursquare points overlaid

We'll save this data into a new GeoJSON file to save preprocessing time if we want to perform analysis on only this subset of data. We first create an output file—in this case, a GeoJSON file. We then use the `.to_file` method to write the GeoJSON file in the output file created:

```
out = r"data/nyc_foursquare.geojson"
nyc_points.to_file(out)
```

Location data visualization

So far, we have been using GeoPandas plotting methods to visualize our location data. GeoPandas is built on top of cartopy, a lower-level Python library, and Geoplot. The plotting functionality in GeoPandas is a very important and convenient way to visualize geographic data; however, it lacks some advanced location data visualization. In this section, I will briefly introduce to you to Folium, which offers an easy-to-use and interactive high-level API for location data visualization.

Folium is a widely used library in the location data visualization ecosystem, as it provides both the mapping flexibility of Python and the visualization strength of the `LeafletJS` library from JavaScript. Folium provides interactive map visualization with different visualization techniques, as well as a number of beautiful base maps. It is also integrated well with the GeoPandas library.

To visualize location data in Folium, you first need to construct a map with a location and, optionally, tiles and a zoom level. This will produce an empty map of the given location (NYC, in our case):

```
import folium

m = folium.Map (
    location = [40.71981037548853, -74.00258103213994],
    #tiles='Mapbox Bright',
    #zoom_start = 11)
```

Once you have constructed your Folium map, you can pass your GeoPandas `GeoDataFrame` and add it to your map (m). This will produce a base map with our `GeoDataFrame` points:

```
folium.GeoJson(nyc_points.sample(1000)).add_to(m)
m
```

The following screenshot is a representation of Folium map with Foursquare points:

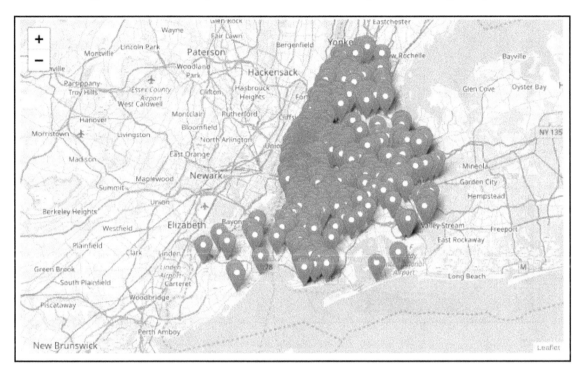

Folium map with Foursquare points

Folium also provides a number of plugins for some advanced visualization techniques, including techniques for clustering points based on their locations. The following code uses the FastMarkerCluster plugin from Folium to do this:

```
from folium.plugins import FastMarkerCluster
sample = nyc_points.sample(10000)
lons = sample['Longtitude']
lats = sample['Latitude']
m = folium.Map(
    location = [np.mean(sample.Latitude), np.mean(sample.Longtitude)],
    tiles= 'Stamen Toner',
    zoom_start=13
    )
FastMarkerCluster(data=list(zip(lats, lons))).add_to(m)
folium.LayerControl().add_to(m)
m
```

The following screenshot shows us when the Foursquare points are clustered in Folium:

Clustered Foursquare points in Folium

Summary

In this chapter, we looked at how to perform spatial operations in a real-world dataset, the Foursquare check-ins. We first created a pandas `DataFrame` from the text file and processed this text file into a `GeoDataFrame`. During this transformation process, we touched on geometries and CRS, as well as geographic coordinate projections. After this, we carried out spatial operations on the `GeoDataFrame`, such as using buffers to calculate distances around subway points and using spatial joins to derive points with NYC district boundaries. Finally, we covered some interactive geographic data visualization techniques using the Folium library in Python.

In the next chapter, we will learn how to aggregate data using machine learning clustering techniques with spatial data. We will use a dataset of reviews about Airbnb properties in Amsterdam to see how to aggregate and cluster points and polygon data using scikit-learn, k-means, and **Density-Based Spatial Clustering Applications with Noise (DBSCAN)** clustering techniques. We will also cover spatial regression and spatial autocorrelation in the next chapter.

4
Making Sense of Humongous Location Datasets

Location data is often complex and contains multiple dimensions that are hard to summarize into a manageable location variable. Geospatial clustering techniques handle these problems by reducing the dimensionality of location data into smaller, manageable, and relevant variables for the data analysis process. Clustering technique importance increases as the amount of data grows.

Location clustering can be referred to as the grouping of different objects into clusters that are similar to each other and fall within the same geographic area. Here, similarity is the metric used to indicate how relationships are strong in different locations.

This chapter tries to explain and explore clustering techniques, as we will use machine learning and spatial statistics to derive an insightful location analysis with less dimensional complexity. We will cover the following topics in this chapter:

- K-means clustering
- **Density-Based Spatial Clustering Applications with Noise (DBSCAN)**
- Spatial autocorrelation

K-means clustering

K-means clustering is one of the most widely used unsupervised machine learning techniques and is used mainly for data mining purposes. K-means is not particularly; exclusive to location data but is also used in diverse applications to partition observations into clusters (k). In a classic k-means clustering, the full weights are on attribute similarity, while location-based k-means specifically targets geographic coordinates to derive spatial or location similarity. We will use the latter as we are interested in location data analysis.

The k-means algorithm is based on randomly selecting k (where k is the number of clusters specified) number of objects that represent initially a cluster mean or center. Then, the algorithm assigns other objects to the cluster, which is closely based on the Euclidean distance between the object and cluster mean. The k-means process is iterative and requires heavy computations when applied to a large dataset as it goes through each object iteratively.

Before we delve deep into k-means clustering, let's have a look at this chapter's dataset. We will use the crime dataset from the Avon and Somerset Constabulary, a territorial police area in the UK.

The crime dataset

The crime dataset we will use for this chapter is from the Avon and Somerset Constabulary. It consists of a number of attributes such as crime ID, month (only February, 2019), and crime type, as well as who reported it. It also has location features such as neighborhood, location, LSOA code, and coordinates.

To read the data, we will use the pandas `.read_csv` method and pass through the name of the dataset:

```
crime_somerset = pd.read_csv("2019-02-avon-and-somerset-street.csv")
crime_somerset.head()
```

These are the first five rows of the dataset. The location features are stored in the `Longitude` and `Latitude` columns:

	Crime ID	Month	Reported by	Falls within	Longitude	Latitude	Location	LSOA code	LSOA name	Crime type	Last outcome category	Context
0	NaN	2019-02	Avon and Somerset Constabulary	Avon and Somerset Constabulary	-2.515072	51.419357	On or near Stockwood Hill	E01014399	Bath and North East Somerset 001A	Anti-social behaviour	NaN	NaN
1	NaN	2019-02	Avon and Somerset Constabulary	Avon and Somerset Constabulary	-2.516919	51.423683	On or near A4175	E01014399	Bath and North East Somerset 001A	Anti-social behaviour	NaN	NaN
2	NaN	2019-02	Avon and Somerset Constabulary	Avon and Somerset Constabulary	-2.516919	51.423683	On or near A4175	E01014399	Bath and North East Somerset 001A	Anti-social behaviour	NaN	NaN
3	NaN	2019-02	Avon and Somerset Constabulary	Avon and Somerset Constabulary	-2.509384	51.409590	On or near Barnard Walk	E01014399	Bath and North East Somerset 001A	Anti-social behaviour	NaN	NaN
4	NaN	2019-02	Avon and Somerset Constabulary	Avon and Somerset Constabulary	-2.509126	51.416137	On or near St Francis Road	E01014399	Bath and North East Somerset 001A	Anti-social behaviour	NaN	NaN

Crime dataset

The shape of this dataset is 13,183 unique crimes reported in the Avon and Somerset Constabulary. As we have seen in `Chapter 3`, *Performing Spatial Operations Like a Pro*, we need to convert this plain CSV into a `GeoDataFrame`. Let's create a function to do so that we can use later for converting any arbitrary CSV dataset with latitude and longitude. But before we do that, let's clean this data before processing it into a `GeoDataFrame`. As you can see, there are some missing values with `NaN` in the preceding table.

Cleaning data

Cleaning data constitutes a major part of doing data science. It is rarely the case to get data in a suitable condition for your application. Therefore, it is necessary to carry out preprocessing and cleaning out data.

Let's check first how many null values we have in our data. We use the pandas `.isnull()` function and `.sum()` to get each column's total null values:

```
crime_somerset.isnull().sum()
```

The output of the preceding code is as follows. Each column name is displayed, followed by the number of missing values in that column:

```
Crime ID 2608
Month 0
Reported by 0
Falls within 0
Longitude 1400
Latitude 1400
Location 0
LSOA code 1400
LSOA name 1400
Crime type 0
Last outcome category 2608
Context 13183
```

The output indicates that there some columns with more 2,000 `NaN` values such as `Crime ID` and `Last outcome category`, while some others have 1,400 missing. We can drop all columns with more than a threshold, for example, 2,000 rows of missing values while maintaining columns with missing values less than the specified number (2,000 in our case).

To drop a column for those with high values of missing rows, we can use the following code to drop the three columns with the highest missing values, `Last outcome category`, `Context`, and `Crime ID`:

```
crime_somerset.drop(['Last outcome category','Context', 'Crime ID' ],
axis=1, inplace=True)
```

Now, those three columns are all dropped from our dataset. We need to drop rows of missing values to clean our data and get it ready for machine learning models. To drop rows with any missing values in the dataset, we can do the following:

```
crime_somerset.dropna(axis=0,inplace=True)
```

If you run this code again, you will see that the whole dataset does not have any missing values:

```
crime_somerset.isnull().sum()
```

Now that we have a clean dataset, let's convert the pandas `DataFrame` into a GeoPandas `GeoDataFrame`.

Converting into a GeoDataFrame

In `Chapter 3`, *Performing Spatial Operations Like a Pro*, we converted a pandas `DataFrame` into a GeoPandas `GeoDataFrame` without creating a function. If you perform the same tasks again and again, creating a function to reuse it becomes very helpful. Here is a function that creates a `GeoDataFrame` using a pandas `DataFrame`. We can use this function to create a `GeoDataFrame` from any CSV file with `Latitude` and `Longitude` columns:

```
def create_gdf(df, lat, lon):
    """ Convert pandas dataframe into a Geopandas GeoDataFrame"""
    crs = {'init': 'epsg:4326'}
    geometry = [Point(xy) for xy in zip(airbnb[lon], airbnb[lat])]
    gdf = gpd.GeoDataFrame(airbnb, crs=crs, geometry=geometry)
    return gdf
```

Now that we have created a function to convert a pandas `DataFrame` into a GeoPandas `GeoDataFrame`, we can use to call that function by providing the names of the coordinates in case the dataset has different names for latitude and longitude. Let's call the function on the crime dataset:

```
crime_somerset_gdf = create_gdf(crime_somerset, 'Latitude', 'Longitude')
```

If you look at the `GeoDataFrame` we have just created, you can see an additional column, which is the `geometry` column. Now, we can plot maps and use other geographic processing and analysis enabled by this additional column. Let's plot and have a look at our dataset:

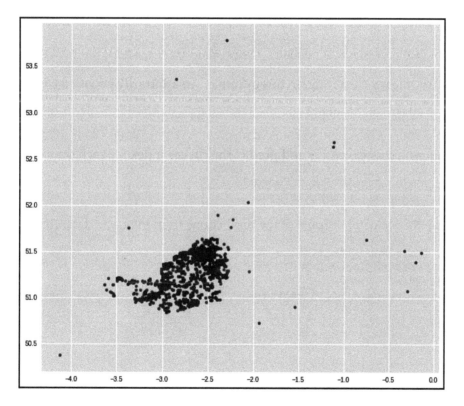

Avon and Somerset crime points

K-means clustering with scikit-learn

To apply k-means clustering on location data, we need to get the coordinates of these features. Before we do that, we will split the dataset into train and test datasets. The test dataset will be used for predicting which group a point belongs to:

```
from sklearn.cluster import KMeans

train = airbnb.sample(frac=0.7, random_state=14)
test = airbnb.drop(train.index)
```

Now that we have created a training and test dataset, let's store training and test coordinates:

```
train_coords = train[['latitude', 'longitude']].values
test_coords = test[['latitude', 'longitude']].values
```

Let's compute k-means clustering. In this example, we arbitrarily choose 5 clusters:

```
kmeans = KMeans(n_clusters=5)
kmeans.fit(train_coords)
```

Then, we compute cluster centers and predict the cluster index for each sample:

```
preds = kmeans.predict(test_coords)
centers = kmeans.cluster_centers_
```

Let's visualize the 5 cluster outputs of the predictions from the test dataset:

```
fig, ax = plt.subplots(figsize=(12,10))
plt.scatter(test_coords[:, 0], test_coords[:, 1], c=preds, s=30,
cmap='viridis')
plt.scatter(centers[:,0], centers[:,1], c='Red', marker="s", s=50);
```

The output of the preceding code is as follows, where each cluster is delineated based on the cluster centers:

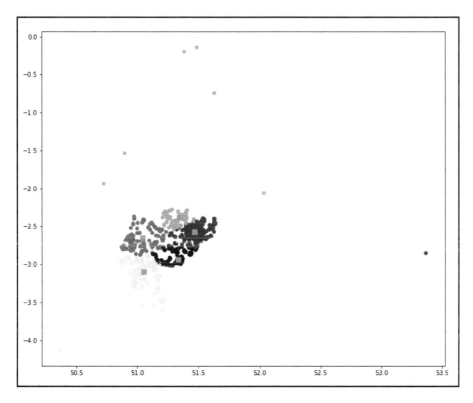

K-means clustering with five clusters

The center points are displayed as squares while other cluster points are displayed as a circle point. Each cluster measures how close it is to that mean. We have some outlier points here at the corners, but they clearly show how the k-means algorithm works in this case. Each point is clustered according to the nearest center mean point.

Density-Based Spatial Clustering Applications with Noise

While k-means clustering relied on providing the number of clusters beforehand, the DBSCAN algorithm is a non-parametric algorithm. Given a set of points, DBSCAN groups together points that are close to each other while also marking outliers. This algorithm can identify clusters even in large spatial datasets by simply highlighting the local density of points. It is also one of the most widely used clustering algorithms, especially for location data. DBSCAN requires two parameters to be supplied before running the algorithm: epsilon and minimum points or samples. Their values significantly influence the results of this algorithm and therefore require some fine-tuning, as well as exploration, before finding suitable clusters.

Epsilon is the parameter that specifies the radius of a neighborhood with respect to other points. In a given set of points, epsilon indicates the distance a point lies closer to a cluster of other points. On the other hand, minimum samples or points set the minimum number of points needed to form a cluster. Based on these two parameters, DBSCAN classifies into three types (this can be more or less depending on these two parameters, but at least one cluster is present):

- **Core points (0)**: A point is a core point when it fulfils the epsilon distance, has the minimum points, and is a neighbor with another core point.
- **Border points (1 to *n*)**: A point is a border point when it does not fulfil the minimum points required but does share a neighborhood with at least one core point. The number of clusters in this category can be many depending on the parameters.
- **Noise Points (-1)**: A point that does not fulfil the epsilon distance, does not have the minimum points required, and does not share any neighborhood with other core points is known as **noise**. These are considered outliers.

We can use DBSCAN to primarily detect outliers or noise or in contrast main clusters. We will cover both of these use cases in the next two sections: *Detecting outliers* and *Detecting clusters*. We will be using the DBSCAN algorithm from the scikit-learn library.

Detecting outliers

Let's first create coordinates out of the latitude and longitude for the whole data, since we do not need to split the train and test dataset for this algorithm:

```
coords = crime_somerset_gdf[['Latitude', 'Longitude']]
```

Now, we are all set to apply DBSCAN on the coordinates. DBSCAN returns tuples; we are only interested in the labels and not the index and therefore we label it a `_` symbol, which denotes generally unwanted results:

```
_, labels = dbscan(crime_somerset_gdf[['Latitude', 'Longitude']], eps=0.1,
min_samples=10)
```

Here, we pass an epsilon of `0.1` and minimum samples of `10` points, a higher epsilon and lower samples to detect outliers. Let's create a `DataFrame` out of the labels result and group according to the clusters:

```
labels_df = pd.DataFrame(labels, index=crime_somerset_gdf.index,
columns=['cluster'])
labels_df.groupby('cluster').size()
```

With these epsilon and minimum sample parameters, DBSCAN results indicate that only 18 points are classified as noise or outliers, while all other points fall into the core cluster. Let's assign each category a name and plot it to see the results.

First, we subset the noise and core from `labels_df` and create a separate `DataFrame` for each:

```
noise = crime_somerset_gdf.loc[labels_df['cluster']==-1, ['Latitude',
'Longitude']]
core = crime_somerset_gdf.loc[labels_df['cluster']== 0, ['Latitude',
'Longitude']]
```

Now, we will plot them using the scatter plot. Here, we use it to display the noise as stars while the core is displayed as circles:

```
fig, ax = plt.subplots(figsize=(12,10))
ax.scatter(noise['Latitude'], noise['Longitude'],marker= '*', s=40,
c='blue' )
ax.scatter(core['Latitude'], core['Longitude'], marker= 'o', s=20, c='red')
plt.show();
```

The scatter plot for the results is as follows. As you can see, the 18 points detected as noise (outliers) are displayed as stars while the core points are displayed as circles:

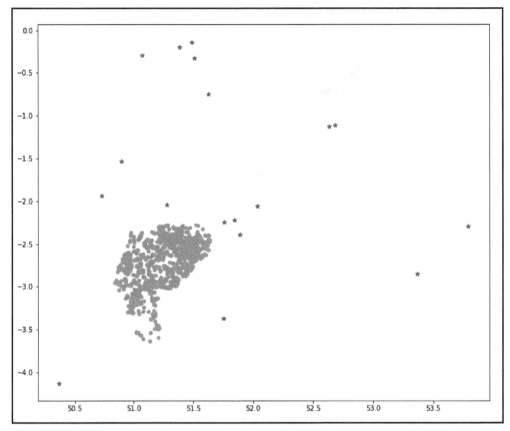

DBSCAN algorithm clusters: outlier detection

We can also use DBSCAN to detect clusters instead of outliers. In the following section, we will cover tweaking this algorithm to detect clusters.

Detecting clusters

To detect clusters, we need to make the epsilon lower while increasing the minimum samples. Let's look at one example of such a scenario. We will set the epsilon as `0.01` and make the minimum samples higher, at `300` points:

```
_, labels = dbscan(crime_somerset_gdf[['Latitude', 'Longitude']], eps=0.01,
min_samples=300)
labels_df = pd.DataFrame(labels, index=crime_somerset_gdf.index,
columns=['cluster'])
labels_df.groupby('cluster').size()
```

In this case, we have more than two clusters as there are some border points (three border point clusters). Let's create a cluster for each one of them and plot them in a scatter plot:

```
noise = crime_somerset_gdf.loc[labels_df['cluster']==-1, ['Latitude',
'Longitude']]
core = crime_somerset_gdf.loc[labels_df['cluster']== 0, ['Latitude',
'Longitude']]
bp1 = crime_somerset_gdf.loc[labels_df['cluster']== 1, ['Latitude',
'Longitude']]
bp2 = crime_somerset_gdf.loc[labels_df['cluster']== 2, ['Latitude',
'Longitude']]
bp3 = crime_somerset_gdf.loc[labels_df['cluster']== 3, ['Latitude',
'Longitude']]
```

Now that we have created each cluster into a separate `DataFrame`, we can plot them using a scatter plot, as follows. To display the image clearly, we also limit the *x* axis and *y* axis to zoom into the clusters:

```
fig, ax = plt.subplots(figsize=(15,12))
ax.scatter(noise['Latitude'], noise['Longitude'],s=1, c='gray' )
ax.scatter(core['Latitude'], core['Longitude'],marker= "*", s=10, c='red')
ax.scatter(bp1['Latitude'], bp1['Longitude'], marker = "v", s=10,
c='yellow')
ax.scatter(bp2['Latitude'], bp2['Longitude'], marker= "P", s=10, c='green')
ax.scatter(bp3['Latitude'], bp3['Longitude'], marker= "d", s=10, c='blue')
ax.set_xlim(left=50.8, right=51.7)
ax.set_ylim(bottom=-3.5, top=-2.0)

plt.show()
```

The output of the scatter plot is shown in the following diagram. There are five clusters detected. The noise (-1) is shown as circles and these are dimmed. The core points are shown with star markers and lie on clustered points at the upper north area of the plot. The three lower clusters are border points:

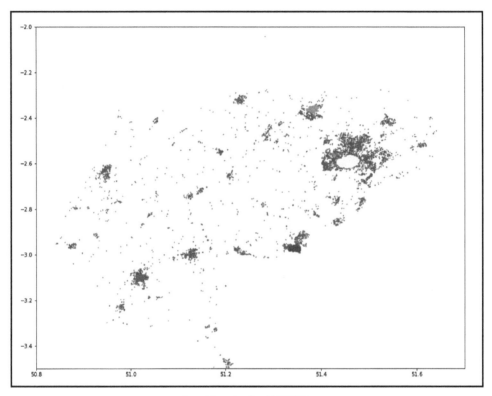

Zoomed cluster detection with DBSCAN

This clearly shows where clusters lie in our data. In the next section, we will cover more elaborate cluster detection techniques using spatial autocorrelation.

Spatial autocorrelation

Spatial autocorrelation is considered an **Exploratory Spatial Data Analysis (ESDA)** method where the concern is to visualize different patterns and clusters through geovisualization and formal statistical tests. Here, the intent is to highlight and explore the similarity of any given value in a dataset to similarity in terms of locations. Therefore, the concept of spatial autocorrelation relates to the combination of similarity between attributions and location.

In contrast to traditional statistical correlations, it does not target the relation of two variables and the change of one value in relation to the other. But spatial autocorrelation focuses on the value of the interested variable in relation to its location and surrounding locations. In other words, spatial autocorrelation allows us to study and understand the spatial distribution and structure of the dataset.

We will use the PySAL library for this section. PySAL 2.0 is closely integrated with GeoPandas and offers an extensive list of spatial statistics functions. We will divide this section into two subsections. First, we will cover global spatial autocorrelation where the focus is on the overall trend and the degree of clustering for the dataset. Questions such as whether there is there a pattern of geographic distribution in this data can be answered through global spatial autocorrelation methods. Second, we will cover local spatial autocorrelation where the intention is to map out the patterns and clusters of the dataset. Here, clusters can be divided into five types:

- **Hotspots**: It indicates positive spatial autocorrelation, that is, high values surrounded by high values
- **Coldspots**: It indicates negative spatial autocorrelation. That means low values surrounded by low values
- **Doughnut**: It is an outlier where high values are surrounded by low values
- **Diamond**: It is an outlier where low values are surrounded by high values
- **Non-significant**: This refers to areas that are not significant at a default pseudo-significance level of 0.05

Before we get into spatial autocorrelation, we need to preprocess the dataset. We have been using points data for this chapter, and to clearly illustrate spatial autocorrelation, we will put points into a polygon. We will first download the boundaries dataset through `wget`, unzip it, and read it with GeoPandas:

```
boundaries = gpd.read_file('ASC_Beats_2016.shp')
```

The boundaries dataset is simply a polygon of neighborhoods in the Avon and Somerset Constabulary. We will derive crime points data and populate this in the neighborhood polygons in the next section.

Points in a polygon

The boundaries data falls within the same geographic extent of the points, so let's overlay and visualize the points and boundaries together:

```
boundaries_4326 = boundaries.to_crs({'init': 'epsg:4326'})
fig, ax = plt.subplots(figsize=(12,10))
boundaries_4326.plot(ax=ax)
crime_somerset_gdf.plot(ax=ax, color='red', markersize=5)
```

The following map gives us an overlaid view of points and boundaries:

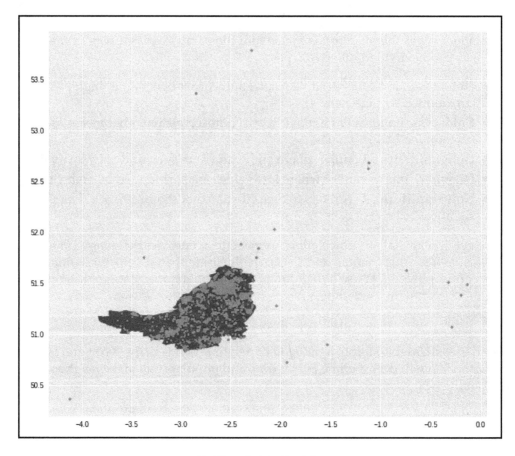

Overlaid map of points and boundaries

We can see that outlier points lie outside of the boundaries. During the spatial join, they will be excluded from the boundaries as they are not within polygon boundaries. Let's perform a spatial join; we have seen one already in Chapter 3, *Performing Spatial Operations Like a Pro*, but the results were points since we used the within operation. To get back a polygon, we can specify operations as contains, and this results in polygons instead of points:

```
crimes_with_boundaries = gpd.sjoin(boundaries_4326,crime_somerset_gdf,
op='contains' )
```

Now that we have merged polygons with points, we still have to do some preprocessing tasks before we can have clean data. Let's look at the first few rows of the result:

	BEAT_CODE	BEAT_NAME	NPA	LPA	PA	geometry	index_right	Month	Reported by	Falls within	Longitude	Latitude	Location	LSOA code	LSOA name	Crime type
0	AE037	Yeovil Centre (One Team)	Yeovil	AE - Somerset East	A - Somerset	POLYGON Z ((-2.627010923440824 50.955264354592...	10230	2019-02	Avon and Somerset Constabulary	Avon and Somerset Constabulary	-2.62479	50.954338	On or near Allingham Road	E01029235	South Somerset 014C	Anti-social behaviour
0	AE037	Yeovil Centre (One Team)	Yeovil	AE - Somerset East	A - Somerset	POLYGON Z ((-2.627010923440824 50.955264354592...	10241	2019-02	Avon and Somerset Constabulary	Avon and Somerset Constabulary	-2.62479	50.954338	On or near Allingham Road	E01029235	South Somerset 014C	Violence and sexual offences
0	AE037	Yeovil Centre (One Team)	Yeovil	AE - Somerset East	A - Somerset	POLYGON Z ((-2.627010923440824 50.955264354592...	10240	2019-02	Avon and Somerset Constabulary	Avon and Somerset Constabulary	-2.62479	50.954338	On or near Allingham Road	E01029235	South Somerset 011C	Violence and sexual offences
0	AE037	Yeovil Centre (One Team)	Yeovil	AE - Somerset East	A - Somerset	POLYGON Z ((-2.627010923440824 50.955264354592...	10239	2019-02	Avon and Somerset Constabulary	Avon and Somerset Constabulary	-2.62479	50.954338	On or near Allingham Road	E01029235	South Somerset 014C	Violence and sexual offences
0	AE037	Yeovil Centre (One Team)	Yeovil	AE - Somerset East	A - Somerset	POLYGON Z ((-2.627010923440824 50.955264354592...	10232	2019-02	Avon and Somerset Constabulary	Avon and Somerset Constabulary	-2.62479	50.954338	On or near Allingham Road	E01029235	South Somerset 014C	Criminal damage and arson

Points in polygon: head

As you can see, the first five rows have same BEAT_CODE. This is not surprising, as there can be many points with one polygon. It happens that all of the first five rows are from BEAT_CODE AE037. We will groupby with BEAT_CODE, get the count of points with each polygon (crimes), and then create a DataFrame. Once this new DataFrame is created, we can merge back as a column to the original boundaries data:

```
grouped_crimes = crimes_with_boundaries.groupby('BEAT_CODE').size()

df = grouped_crimes.to_frame().reset_index()
df.columns = ['BEAT_CODE', 'CrimeCount']

final_result = boundaries.merge(df, on='BEAT_CODE')
```

Now, we have original boundary polygons with an extra column for the number of crimes for each polygon in the boundary dataset. This is how it looks with the first five rows. Pay attention to the last column, where we have crime counts from the points data:

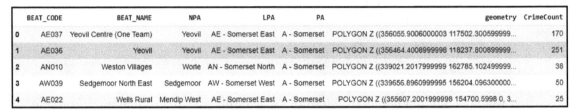

	BEAT_CODE	BEAT_NAME	NPA	LPA	PA	geometry	CrimeCount
0	AE037	Yeovil Centre (One Team)	Yeovil	AE - Somerset East	A - Somerset	POLYGON Z ((356055.9006000003 117502.300599999...	170
1	AE036	Yeovil	Yeovil	AE - Somerset East	A - Somerset	POLYGON Z ((356464.4008999998 118237.800899999...	251
2	AN010	Weston Villages	Worle	AN - Somerset North	A - Somerset	POLYGON Z ((339021.2017999999 162785.102499999...	38
3	AW039	Sedgemoor North East	Sedgemoor	AW - Somerset West	A - Somerset	POLYGON Z ((339656.8960999995 156204.096300000...	50
4	AE022	Wells Rural	Mendip West	AE - Somerset East	A - Somerset	POLYGON Z ((355607.2001999998 154700.5998 0, 3...	25

Crime counts with the boundaries polygon data as columns

Global spatial autocorrelation

As we have seen, global spatial autocorrelation indicates the degree of clustering in a dataset. In other words, the focus is on detecting spatial similarities. The human eye is deceptive when it comes to recognizing and mapping out patterns and clusters.

The choropleth map

Before we start doing spatial autocorrelation, let's first map out a choropleth map for the boundaries according to the number of crimes for each polygon:

```
fig, ax = plt.subplots(figsize=(12,10))
final_result.plot(column='CrimeCount', scheme='Quantiles', k=5,
cmap='YlGnBu', legend=True, ax=ax);
plt.tight_layout()
ax.set_axis_off()
plt.savefig('choroplethmap.png')
plt.title('Crimes Choropleth Map ')
plt.show()
```

The following choropleth map gives an overview of the crime rate:

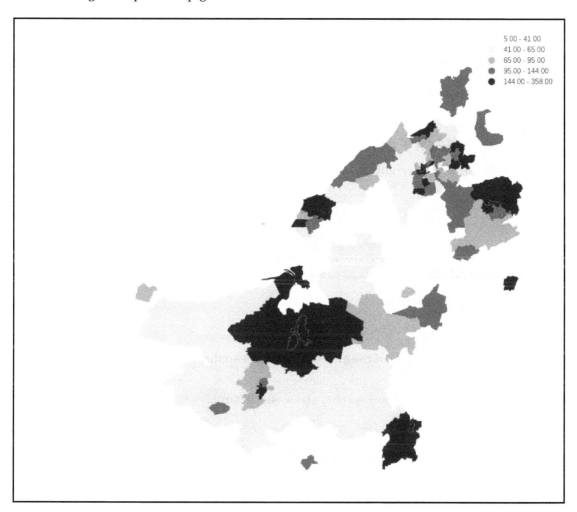

Crimes choropleth map

Although our eyes can detect some high and low crime rate values from the preceding map, it is not easy to point out a pattern of clusters from here. Our eyes might be drawn to look at the darker blue areas that have higher crime rates or light yellow areas with lower crime rate, but how are we sure we are not detecting false positives, or even whether we have a pattern in our data? Here is where we use global spatial autocorrelation to test out whether there are patterns in our data and to what level the pattern of clusters exists. Let's first construct spatial similarity of the dataset and set weights for the spatial similarity.

Spatial similarity and spatial weights

Here, we will use queen weights to construct the spatial similarity of this dataset with PySAL:

```
wq = weights.Queen.from_dataframe(final_result)
wq.transform = 'r'
```

We then apply the weights to the crime counts variable and add the spatial similarity as a column to our final results GeoDataFrame:

```
y = final_result['CrimeCount']
ylag = weights.lag_spatial(wq, y)
final_result['ylag'] = ylag
```

Let's map out the choropleth of the crime data with a spatial similarity:

```
fig, ax = plt.subplots(figsize=(12,10))
final_result.plot(column='ylag', scheme='Quantiles', k=5, cmap='YlGnBu',
legend=True, ax=ax);
plt.tight_layout()
ax.set_axis_off()
plt.savefig('choroplethmap-ylag.png')
plt.title('Crimes Ylag Choropleth Map ')
plt.show()
```

The choropleth map here shows the data classified into a five-quantile scheme:

Spatial similarity crimes choropleth map

If you compare the spatial similarity crimes choropleth map with the previous crimes choropleth map, you will see that the spatial similarity choropleth map is much more smoothed out than the other, and patterns of clusters become more apparent. We have much more smoothed out lower crime rates in the middle while, on the east and west, we have some spatial concentration of high crime rates. However, we still have challenges to visually associate a polygon with the value of crime rates. That is, we still cannot prove there are cluster patterns, and where those clusters are. Now, let's carry out the global spatial autocorrelation tests to determine the degree of cluster patterns with this dataset.

Global spatial autocorrelation

There are different techniques to assess global spatial autocorrelation. One of the widely used statistical tests is the Moran's I test, which we can simply perform with the PySAL Python library. The Moran can tell us whether the data is random or whether any geographical pattern is present in the data. We will first use the numerical result of the Moran's I and will interpret the results. Next, we will visualize the result of the statistical test in the Moran's plot, which is a great way to visualize the general geographic patterns present within the data:

```
w = Queen.from_dataframe(final_result)
moran = Moran(y, wq)
moran.I
```

The Moran statistical test needs both the variable we are interested in (y: crime counts) and the spatial weights we have created earlier. Once we provide those two parameters, Moran's I statistical test returns one number, which can give us an indication of whether the spatial distribution is random or not. For this particular dataset, the results show 0.256. This number is a summary of the overall pattern of clustering within our dataset.

This Moran's I value (0.256) indicates that spatial clusters within this dataset are not actually random, but rather there is a significant spatial association between crimes that exist. To complement the interpretation of the Moran's I value, we can also look at the p-value associated with it:

```
moran.p_sim
```

In fact, the p-value of this dataset is very small (0.001), and therefore we can conclude and confirm that a spatial pattern exists in this data and reject the null hypothesis of this data being spatially random.

A good way to visualize and complement the result of Moran's I is to plot the Moran's I plot, where we plot the variable of interest against the spatial lag. We can do that with PySAL, which has a nice and convenient function for this task:

```
plot_moran(moran, zstandard=True, figsize=(10,4))
plt.show()
```

On the left side, we have a reference distribution. On the right side, the Moran scatter plot is shown, where the data is divided into four categories:

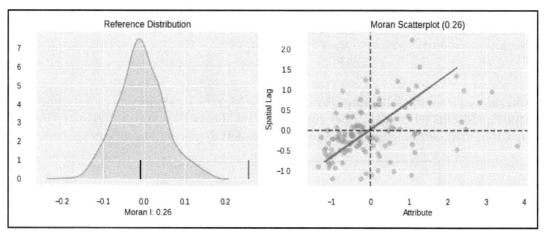

Moran's I plot

The plot clearly displays the distribution of the crimes against the spatial lag as a scatter plot. As you can see from the regression line, there is a positive relationship between the spatial lag and crime rates. This is closely tied to the positive Moran's I value that we have seen earlier as also indicated in the plots. In conclusion, the global spatial autocorrelation focuses only on determining the overall trend; therefore, we can only determine from the Moran's I plots and statistics that there is a positive relationship between the two variables. That means we can observe that there are high crime values close to other high-value crime areas, and vice versa. But we cannot determine where those clusters appear in our data, and that is derived through local spatial autocorrelation in the next subsection.

Local spatial autocorrelation

While the Moran's I test indicated spatial clustering patter present in the data, it did not show us where those patterns exist. With local spatial autocorrelation, we can classify observations into four groups: **high-high (HH)** values near to each other, **low-low (LL)** values, **high-low (HL)** values, and finally, **low-high (LH)** values near to each other:

```
from splot.esda import moran_scatterplot
from esda.moran import Moran_Local
```

In the preceding code, we import `splot` functionality to calculate Local Moran Statistics and display it as a scatter plot. The following code calculates the Local Moran of the `y` variable, and the weights. Then, we use `moran_scatterplot` to plot it:

```
# calculate Moran_Local and plot
moran_loc = Moran_Local(y, wq)
fig, ax = moran_scatterplot(moran_loc)
plt.show()
```

As you can see from the following scatter plot, the Morans local plot can be subdivided it into four quadrants where the upper-right quadrant shows HH groups, the lower-right HL groups, the lower-left LL clusters, while the upper-left displays LH groups:

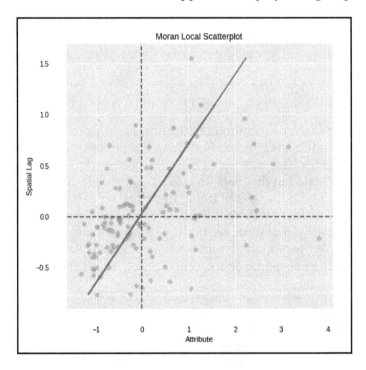

Moran's local scatterplot

More clearly, we can visualize the local spatial autocorrelation as a map with different colors for each cluster:

```
from splot.esda import lisa_cluster
lisa_cluster(moran_loc, final_result, p=0.05, figsize = (10,8))
plt.tight_layout()
plt.show()
```

The following map displays the results of crime clusters in spatial autocorrelation. Now, we can clearly determine hot spot areas with red colors, where high crime rates are clustered together. The blue color indicates cold spots, where, on the contrary, we have lower crime rates close to each other. This map is much more intuitive and clearly portrays the clusters more than the choropleth map we started:

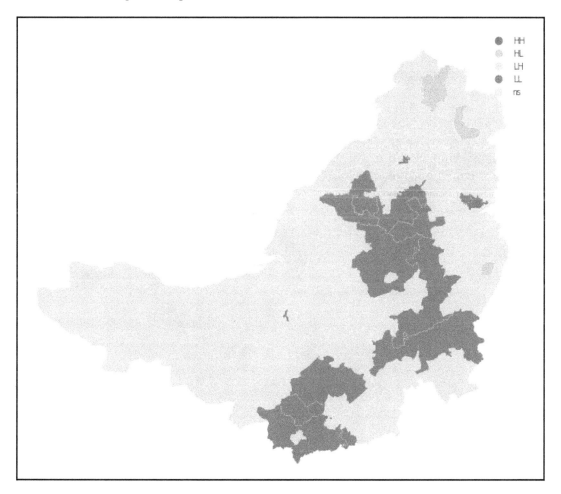

LISA clusters

Global and local spatial autocorrelation statistics are essential techniques to detect clusters in the dataset, in addition to K-means clustering and DBSCAN algorithms.

Summary

In this chapter, we studied different ways to summarize, interpret, and make sense of location data by using machine learning algorithms and spatial statistical methods. We first covered the k-means clustering algorithm, where we created spatial clusters using the scikit-learn library. We then moved on to explore the DBSCAN algorithm to detect outliers as well as clusters. Finally, we studied the two methods of spatial autocorrelation using the PySAL library. Here, we interpreted, plotted, and tested global patterns of the crime dataset. Furthermore, we studied how to derive meaningful and intuitive clusters from the dataset using local spatial autocorrelation.

In the next chapter, we will learn geofencing. Geofences is a popular tool used by businesses as well as conservationists. Geofencing refers to abstract fences that are created around a location, so that an alerting or notification system would notify in case an event happens at or within the fence. The event can be something as simple as a customer entering the vicinity of a business location, or customers within a cell-phone tower range; the applications are unlimited.

5
Nudging Check-Ins with Geofences

Geofencing is a useful tool and concept in geographic data science and is used in many domain applications, including business, conservation, and security, as well as home automation. The concept of geofencing is very simple, yet it is a powerful technique that enriches location applications. Simply put, a geofence is a defining virtual boundary around geographic objects or an area, so that every time a user enters or leaves the boundary perimeters, actions or notifications can be triggered. With the increased use of smartphone, GPS, and location services, geofencing becomes an indispensable tool in location data analytics and intelligence. In this chapter, we will cover the following topics:

- Geofencing concept and components
- Revisiting geometry and topology (lines and polygons)
- Geofencing application example
- Geofencing and interactive web mapping

Geofencing

A geofence involves creating virtual fences around real-world geographic entities. It can be either a dynamically created geofence or predefined boundaries. In a dynamic geofence, the boundaries are created based on a chosen point location and can be created on the fly. This depends on, for example, the current location of a user. In contrast, a static geofence is when predetermined boundaries are used, for example, danger zones, city centers, or a parole exclusionary zone.

Geofencing, on the other hand, contains a geofence and some other components including GPS locations, geometry topologies, and notification systems and entails the use of these components to a certain application. Here, the interaction and use of these components are what enables geofence concepts to be used in wide domain applications. As precise GPS locations with smartphones become ubiquitous, we witness increased use of geofencing applications.

Geofencing applications

Geofencing applications are numerous, as we have seen in the introduction. They can be used in security, business, and smart home design, as well as a ton of other applications. However, the most widely used and common application for geofencing is marketing. In this section, we will cover the use of geofencing in marketing location-based services and products to illustrate and elaborate geofencing concepts with a real-world and concrete example.

Marketing and geofencing

Advertising and marketing agencies have embraced geofencing to increase the effectiveness of their ads and reach their target users efficiently. Marketers deploy geofencing to send or display ads to users at the right time and in the right location once a user is within the perimeters of the geofence, entering it or leaving it. Although this is more likely to be in a mobile device, it also can be effectively used via desktops or tablets, triangulating from Wi-Fi locations. Let's visualize an example of two users:

Geofence: place holder

In this example, **User 1** is in the **Geofence** and, in our hypothetical example, the advertising agency receives an automatic notification once the user has entered the geofence parameters. **User 2** is still not in the **Geofence** parameters in this example. We will implement an example of this in code soon, but before that, we need to revisit geometries, especially lines and polygon geometry. In `Chapter 3`, *Performing Spatial Operations Like a Pro*, we covered only point geometry.

Geometry and topology (lines and polygons)

In this chapter, we will use the Brazil GPS Trajectories dataset from UCI. It contains GPS points (`latitude` and `longitude`) with timestamps. It also contains unique `track_id` for each trajectory.

Here is what the first five rows of the data look like:

	id	latitude	longitude	track_id	time
0	1	-10.939341	-37.062742	1	2014-09-13 07:24:32
1	2	-10.939341	-37.062742	1	2014-09-13 07:24:37
2	3	-10.939324	-37.062765	1	2014-09-13 07:24:42
3	4	-10.939211	-37.062843	1	2014-09-13 07:24:47
4	5	-10.938939	-37.062879	1	2014-09-13 07:24:53

GPS trajectory: first five rows

We create a `GeoDataFrame` using a function we created in `Chapter 4`, *Making Sense of Humongous Location Datasets*:

```
def create_gdf(df, lat, lon):
    """ Convert pandas dataframe into a Geopandas GeoDataFrame"""
    crs = {'init': 'epsg:4326'}
    geometry = [Point(xy) for xy in zip(df[lon], df[lat])]
    gdf = gpd.GeoDataFrame(df, crs=crs, geometry=geometry)
    return gdf

traj_gdf =  create_gdf(trajectories, "latitude", "longitude")
aracaju_city_points = traj_gdf[(traj_gdf['latitude']<-10.80) &
(traj_gdf['longitude']>-37.5)]
```

In this function, we have created a trajectory `GeoDataFrame`, and we then took a subset of the trajectories for the city of Aracaju. This is the main file we will use for this chapter. The trajectories map can be displayed in a simple map with GeoPandas using `.plot()`:

```
fig, ax = plt.subplots(figsize=(8,8))
aracaju_city_points.plot(markersize=5, cmap='Dark2', ax=ax)
plt.tight_layout()
plt.axis('off')
plt.savefig('aracaju_trajectory.png')
plt.show()
```

This plot shows the trajectory points of `aracaju_city_points`:

Aracaju city trajectory GPS points

As you can see, these trajectory points are dispersed and separated but clearly show a trajectory line. In the next section, we will cover how to convert point geometries into line geometries.

Line geometries

To create a line geometry, we need to have at least two points, and it is almost similar to how we created points in Shapely. Let's create a line from `track_id_1`. First, we subset the data to get only this track ID, and then we use the Shapely geometry library to create the line:

```
track_id_1 = aracaju_city_points[aracaju_city_points['track_id']== 1]

line = LineString(track_id_1.geometry)
```

Now that you have seen how to create one line, let's do this for the whole dataset. Remember, we said, to create line geometry, we need to filter out and take only the track IDs with two or more points before we can create a line geometry for the whole dataset:

```
filtered_trackid = aracaju_city_points.groupby('track_id').filter(lambda x: len(x) >= 2)
```

Then, we group all `filtered_trackid` instances on the `track_id` feature, get the geometry of the points in each `track_id`, and apply a Shapely `LineString` geometry on points in each `track_id`. This is a chained process, but you can take each part and reconstruct it:

```
all_tracks =
filtered_trackid.groupby(['track_id'])['geometry'].apply(lambda x:
LineString(x.tolist()))
```

In return, we get all of the lines converted out into a `LineString` geometry. If we want to create a `GeoDataFrame` out of this, the procedure is the same as we covered in Chapter 3, *Performing Spatial Operations Like a Pro*, and Chapter 4, *Making Sense of Humongous Location Datasets*:

```
gdf_lines_all = gpd.GeoDataFrame(all_tracks, geometry='geometry', crs = {'init':'epsg:4326'})

gdf_lines_all['track_id'] = gdf_lines_all.index
gdf_lines_all.reset_index(drop=True, inplace=True)
```

Now, we have a `GeoDataFrame` with a `LineString` geometry for all track IDs. Let's have a look at the `LineString` map:

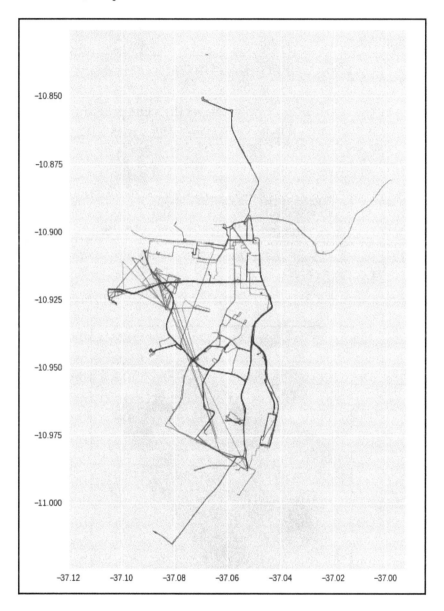

Trajectory points converted into lines geometry

Polygon geometries

To create a polygon, you need to have at least three coordinates, so it can form a triangle polygon. Let's create a polygon using Python and Shapely. In this case, I grabbed four coordinates around Aracaju city, Brazil:

```
lats = [-10.813777, -11.002150, -11.070560,-10.878416]
lons = [-37.079790, -37.203427, -37.109280, -36.986931]
```

To create a polygon, `GeoDataFrame` simply follows the same procedure we have used to create a point and line geometry. We need to first create the geometry out of the coordinates as well as the coordinate reference system:

```
crs = {'init': 'epsg:4326'}

polygon_geometry = Polygon(zip(lons, lats))
```

Now, we can create `GeoDataFrame` out of the geometry and CRS:

```
polygon_gdf = gpd.GeoDataFrame(index=[0], crs=crs,
geometry=[polygon_geometry])
```

Let's visualize and overlay the polygon and the lines we have just created:

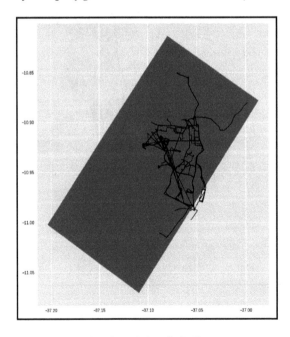

Overlayed lines and polygon GeoDataFrames

Topology – points in a polygon

We already covered points in a polygon with `sjoin` back in `Chapter 3`, *Performing Spatial Operations Like a Pro*, and will revisit it here as a topology operation. Pay attention to this example, as this is the most important aspect of the geofencing concept. To simplify our case, we will first take a buffered polygon out of one trajectory point. Later, when doing geofencing, we will bring out polygons of geographic objects, such as an airport, beach, and city center. We convert the buffered point into a polygon `GeoDataFrame`, reset the index, and make sure that the name of the geometry column is `geometry`:

```
# Create buffer on Point 20 in track_id 1
buffer = track_id_1[track_id_1['id'] == 20].buffer(0.005)
buffer.reset_index(drop=True, inplace=True)
buffer_gdf = gpd.GeoDataFrame(buffer)
buffer_gdf.columns = ['geometry']
```

Let's visualize the tracking points on top of the buffer polygon:

```
# Plot track_id 1 points over the Buffer Polygon
fig, ax = plt.subplots(figsize=(10,10))
buffer_gdf.plot(ax=ax)
track_id_1.plot(ax=ax, color='black')
plt.tight_layout()
#plt.savefig('polygon_lines.png')
#plt.axis('off')
plt.show()
```

The following map shows the buffered polygon overlaid with `track_id_1` points. As you can see, some points are inside `buffer_gdf` while others are outside of the polygon:

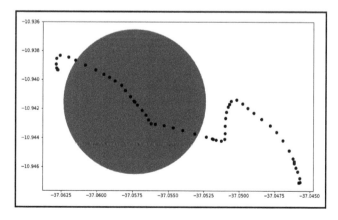

Overlay track_id_1 and buffer_gdf

As you can see, some points fall inside the buffer polygon while others are outside the polygon. We will use the topology operation (`within`) to check each point's position:

```
# Mask points in polygon -> returns True if inside polygon
pip_mask = track_id_1.within(buffer_gdf.loc[0, 'geometry'])
pip_mask.value_counts()
```

`pip_mask` returns a series of `True` or `False` instances to indicate whether a point is inside the polygon or not. We can count out the values of this series. In this case, 62 points fall outside the polygon, while 28 points fall inside the polygon. Let's loop over the first 12 values of `pip_mask` and print out whether it is inside the buffer or outside:

```
for i in pip_mask[:12]:
 if i == True:
  print('Inside the Buffer')
 else:
  print("Otuside the Buffer")
```

The preceding prints out a `Outside the Buffer` or `Inside the Buffer` message for each point, as shown as follows:

```
Outside the Buffer.
 Outside the Buffer.
 Outside the Buffer.
 Outside the Buffer.
 Outside the Buffer.
 Outside the Buffer.
 Outside the Buffer.
 Outside the Buffer.
 Outside the Buffer.
 Inside the Buffer.
 Inside the Buffer.
 Inside the Buffer.
```

Let's add these mask values to our `track_id_1`, variable of `GeoDataFrame` so that we can check it when plotting. In this example, once we filter out `track_id_1` we can differentiate between the points that are inside the polygon or outside the polygon as per the following plot. Here, points have different color, depending on their position:

```
# Plot track_id 1 points over the Buffer Polygon
fig, ax = plt.subplots(figsize=(10,10))
buffer_gdf.plot(ax=ax)
track_id_1.plot(ax=ax, column='pip_mask', cmap='bwr')
plt.tight_layout()
plt.savefig('pipmask_buffer.png')
#plt.axis('off')
plt.show()
track_id_1.loc[:,'pip_mask'] = pip_mask
```

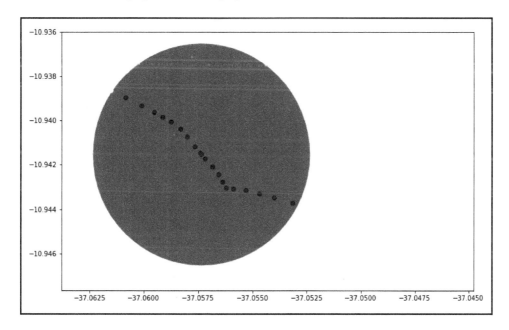

Masked points in the buffer (red)

Geofencing with Plotly

Now that we have covered all of the components of geofencing, we can perform geofencing concepts in practice using our trajectory data. We will utilize the Plotly visualization library to animate trajectories and visualize our geofencing application. We will bring three polygons that denote different use cases: an airport, beach, and city center.

Masking

First, we need to create a mask where we store whether a certain point is inside the geofence or not. We first read the `geofence_polygons` from the dataset provided with Notebook. Upload it first in Google Colab:

```
geofence_polygons = gpd.read_file('geofence_polygons.gpkg')
geofence_polygons
```

	id	name	geometry
0	1	Airport	(POLYGON ((-37.08070101133483 -10.990966514049...
1	2	Beach	(POLYGON ((-37.05289499526379 -10.985695231382...
2	3	Center	(POLYGON ((-37.05481681706964 -10.906043953908...

Geofence polygons

Let's plot and overlay `geofence_polygons` and `filtered_trackid` points on top of each other:

```
# Plot filtered_tracid points over geofence polygons
fig, ax = plt.subplots(figsize=(10,10))
geofence_polygons.plot(ax=ax, color='gray')
filtered_trackid.plot(ax=ax, markersize=1)
plt.tight_layout()
#plt.savefig('polygon_lines.png')
#plt.axis('off')
plt.show()
```

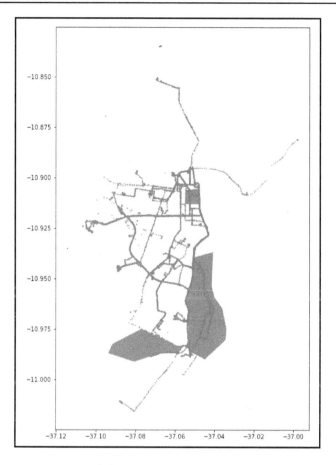

Geofence polygons (gray) and tracking points

Let's perform masking based on these three polygons in `geofence_polygons`. Once we do that, we filter out all points with the mask using the GeoPandas `within` operation:

```
mask = (geofence_polygons.loc[0, 'geometry'])| (geofence_polygons.loc[1,
'geometry']) |(geofence_polygons.loc[2, 'geometry'])


pip_mask_qeofence = filtered_trackid.within(mask)
pip_mask_geofence.value_counts()
```

The value counts of tracking points within these three polygons are 2,257, while the other 14,420 are outside of the geofence polygons.

Plotly interactive maps

Plotly is an analytics web application framework that provides interactive tools such as graphs and maps, and analytics and statistics tools, in the web. Recently, they have released Plotly Express, a simple interactive consistent high-level API for Python visualization. We will use Plotly Express to visualize our geofencing application. It will simply animate points, color differently when the points are inside the geofence, and give us a legend of whether a point is inside or outside of the geofence polygon.

 You need a Mapbox token to use the functionality of `scatter_mapbox`. Go to `https://www.mapbox.com/` and register; it is free. Then, you need to go to **Account** | **Tokens** | **Create Tokens**. Copy and paste your token to this function before using Plotly Express `scatter_mapbox`.

Let us import Plotly express and set up Mapbox tokens:

```
import plotly_express as px
px.set_mapbox_access_token("PASTE TOKEN HERE")
```

Let's first visualize a sample of tracking points with Plotly Express. We will use the `scatter_mapbox` function, which takes the data, `latitude` and `longitude`, and some other optional fields such as `color` and `size`:

```
px.scatter_mapbox(filtered_trackid.sample(500),
                  lat="latitude",
                  lon="longitude",
                  color="geofence",
                  size='track_id' ,
                  size_max=12,
                  zoom=12
)
```

 Visualizing the complete tracking points might take a long time, and your Google Colab might crash.

Visualizing the complete tracking points might take a long time, and your Google Colab might not work. Therefore, we only visualize a sample in the preceding code:

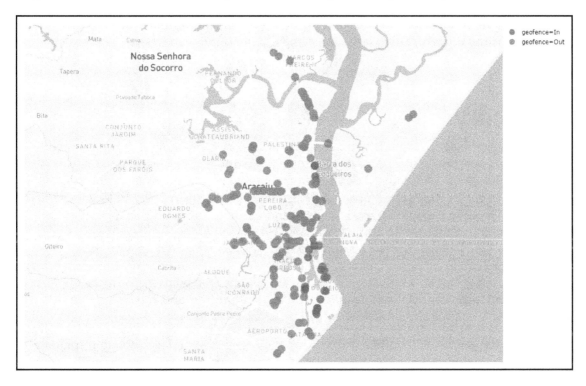

Plotly Express scatter plot for the trajectory data (color by geofence in or out)

Now that we have laid out all of the foundational work for the geofence application, the last step is to animate the points and see how the geofence works in a real-world application. `scatter_mapbox` has an argument parameter for that, and we only need to provide the time frame; in our case, we have a time column:

```
px.scatter_mapbox(filtered_trackid[:100],
  lat="latitude",
  lon="longitude",
  color="geofence",
  size='track_id',
  animation_frame='time',
  size_max=15, zoom=12)
```

You can see the final map in the Google Colab Notebook or any other Jupyter Notebook you are using. It is pretty advanced visualization with a little code that you can start experimenting with right away. Plotly Express works perfectly well with Plotly, which provides extensive tools for visualization including online and offline charts, graphs, and maps. For example, with Plotly Dash, you can build beautiful web-based analytics applications.

Summary

In this chapter, we covered geofencing concepts and what geofencing is and presented different use cases and applications of geofencing. We touched upon and revisited geometry operations, especially `LineString` and polygons. To illustrate a simple application of geofencing, we first covered masking with points in polygon topology operations. Once we mastered these geometry and topology concepts, we moved on to using geofencing in real-world application data. Putting all of these different pieces together, we were able to construct a mask for the whole trajectory data with three geofencing polygons: an airport, beach, and city center. Finally, we animated the trajectory points, where trajectory points move and indicate whether the point is inside the geofence polygon or outside. The applications of geofencing are limitless and can be applied in different case scenarios and domains.

In the next chapter, we will build a routing engine, where we will learn about graph data structures and network analysis. We will carry out network analysis, shortest path cost, routing, and isochrones with real-world data using **Open Street Map** (**OSM**) data. We will utilize the NetworkX library with pandas and GeoPandas libraries to process the geographic data.

Let's Build a Routing Engine 6

"Logic will take you from A to B. Imagination will take you everywhere."

- Albert Einstein

Despite the possibility of flying cars in the near future, right now, you still need to use the road or rail to get from point *A* to point *B* on land. I am pretty sure that you have never once deleted the *Maps* app from your smartphone. So, what makes the *Maps* app so indispensable that you can't imagine living in the pre-Google Maps era (that is, pre-2005)? Map-based routing saves you thousands of dollars in fuel costs and time spent in traffic (unless you're in the Bay Area, in which case, even Google Maps can't save you from the traffic!).

Good map-based routing is dependent on an accurate, well-defined, and updated graph network. Graph algorithms are treated as an advanced topic in computer science since few of the graph problems such as the **traveling salesman problem** (**TSP**) are considered NP-complete problems. But who says we have to flirt with the NP-completeness paradigm to build a simple routing engine? With open source road traffic data and public transit feeds, routing engines will no longer be a black box.

The following topics will be covered in this chapter:

- Fundamentals of graph data structure
- Shortest path analysis on a simple graph
- Building a graph based on a road network
- Shortest path analysis on the road network graph

 We will be using Google Colab for this chapter. If you have a cloud machine where you've hosted a Jupyter Notebook, that works fine as well.

Fundamentals of graph data structure

Graphs can be effectively used to model and solve routing problems through road and public transit networks. Graphs can be designed to model and predict financial transactions and even complex social networks (yeah, blame a graph algorithm the next time Facebook or LinkedIn makes an unfamiliar or unsolicited friend suggestion or professional connection). Despite its versatility, the graph universe is made up of just two simple, easily relatable components, namely, **nodes** and **edges**. In a road network, a node might represent a road intersection and an edge might very well represent the road segment itself. The convention is that an edge is an entity that always connects two nodes, as is represented in the following diagram:

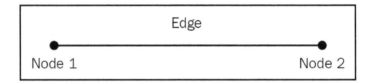

Simple edge

Let's fire up a new Google Colab Notebook and build our first graph using a Python library known as `networkx`. By default, the `networkx` library is installed in the Google Colab environment. If not, be sure to install it using `pip` or `conda` or a similar package manager. The following command should work in most Jupyter Notebook coding environments:

```
!pip install networkx
```

The `networkx` library provides a clean and efficient data structure to define and work with graphs. The simplicity of the `networkx` library is the most appealing factor for adopting this library for this chapter. Let's get started with `networkx` and create our first graph with just four lines of code:

```
import networkx as nx
G = nx.Graph()
G.add_edge('A','B')
G.add_edge('B','C')
```

Plotting this simple graph looks like the following diagram:

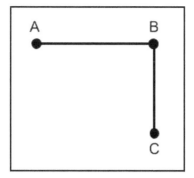

An elementary graph with nodes and edges

Just like that, we have defined a graph object, G, and we have added two edges to it. The first edge connects the 'A' and 'B' nodes, and the second edge connects nodes 'B' and 'C'.

Directional graphs

These edges in the preceding graph didn't have a direction. But edges can have a direction. If you think of the road network analogy, there are one-way roads, in which you can only drive along in one direction, but most roads are bidirectional:

One way roads

This can be modeled in `networkx` by instantiating a **directional graph (digraph)**. In a digraph, the position of the nodes in the *edge definition* determines the direction of the edge. The convention is the first node in the edge definition is the *from node* or *source node* and the second node is the *to node* or *target node* or *destination node*. Let's instantiate a directional graph with the following code snippet:

```
H = nx.DiGraph()
H.add_edge('B', 'A')
H.add_edge('B','C')
```

Plotting the digraph, `H`, yields the following plot. Notice the arrows in the diagram indicating the direction of the edge:

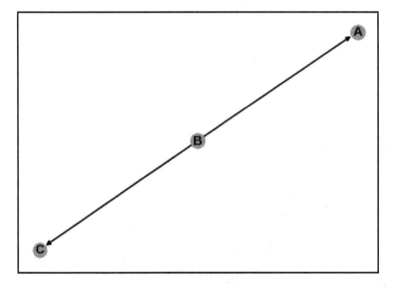

A directed graph

The preceding lines of code mean that there are three nodes, `'A'`, `'B'`, and `'C'`, and there's a connection between B and A and not the other way around. It also means that there's a connection between B and C, but not between C to B. Digraphs are very important in modeling real-world networks, especially road networks.

If a road segment is bidirectional, you might have to add two different edges between the same nodes, as follows:

```
I = nx.DiGraph()
I.add_edge("A", "B")
I.add_edge("B", "A")
I.add_edge("A", "C")
```

In the following plot, notice the arrows. Edge `AB` has two arrows pointing in opposite directions:

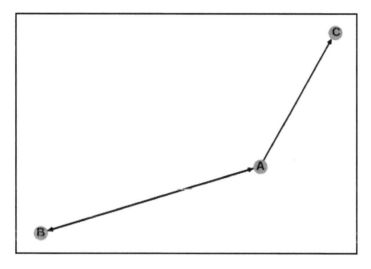

A bidirectional graph

Weighted graphs

In the previous examples, each edge is considered to have a unit weight. What that means is that the cost of traveling from one node to another through an edge is the same. This need not always be the case. In the case of a road network, each road segment is different from each other in terms of length and time (taken to traverse it). So, if we are going to represent these road segments as edges, we need to make sure that the edges have different costs or *weights*.

The `networkx` library allows us to add weight to an edge, which is demonstrated in the code, as follows:

```
import networkx as nx

#Create a weighted graph
G=nx.DiGraph()
G.add_edge('A','B',weight=6)
G.add_edge('A','C',weight=2)
G.add_edge('C','D',weight=4.5)
G.add_edge('C','E',weight=5)
G.add_edge('C','F',weight=6)
G.add_edge('A','D',weight=3)
```

The output will be as follows:

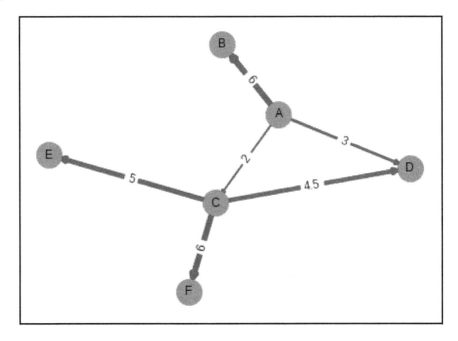

A weighted graph with the width of edges representing the weight of the edge

In this section, we had a good introduction to the components and types of graph data structures. In the next section, we will look into the popular analyses commonly performed using graphs.

Shortest path analysis on a simple graph

Suppose you want to connect to Barack Obama through LinkedIn; how many degrees of connection do you have to go through to reach Obama? A first-degree connection is someone who is connected to you on LinkedIn. A second-degree connection is someone who is connected to your first-degree connection and so on. Assuming that each of your connections and their connections respectively are interested and ready to help you network with the Obama, research says that it only takes an average of 5 degrees of connection for you to connect with Obama, or anyone in the world, for that matter. In other words, the shortest path between any of us and Obama is less than or equal to five. That sounds strange, right? This is known as **the small-world phenomenon**. And, fortunately, the shortest path between **This author** and Obama is only three.

This means that someone **This author** knows on LinkedIn knows someone else who is connected to Obama:

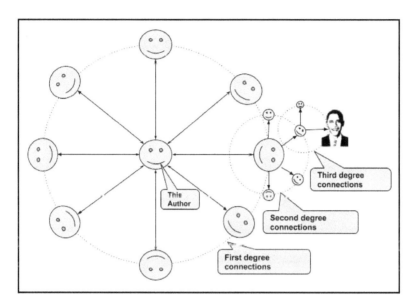

Now, **This author** didn't know that Obama is a third-degree connection until he looked up this information on LinkedIn before writing this. But how did LinkedIn compute this degree of separation? LinkedIn knew this fact because it is continuously running the shortest path algorithm for each member on its network and is able to calculate almost in realtime that the shortest path between **This author** and Obama is three (connections).

Let's look at some code to understand this. Let's create a hypothetical network of friends, centered around a person called JK and link any two individuals in the network to each other with an edge connection if they are friends:

```
L = nx.Graph()
L.add_edge("JK", "Ashish")
L.add_edge("JK", "Athulya")
L.add_edge("JK", "Peter")
L.add_edge("JK", "Eric")
L.add_edge("Peter", "Derrick")
L.add_edge("Peter", "Vijay")
L.add_edge("Eric", "Enoch")
L.add_edge("Athulya", "Jeremiah")
L.add_edge("Derrick", "Obama" )
L.add_edge("Jeremiah", "Aaron")
L.add_edge("Aaron", "Obama")
```

When represented as a diagram, the friend network of JK looks like this:

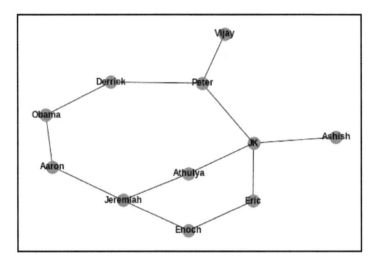

JK's friend network

Now, let's see who can connect JK to Obama. Well, Athulya can connect JK to Obama through Jeremiah, and Jeremiah through Aaron. But voila, Peter can connect JK to Obama through his friend Derrick. Hence, the shortest path for JK to Obama is through Peter and Derrick, respectively, and the shortest path length for JK to Obama is three. This can easily be found using the shortest_path_lenth() method in networkx:

```
nx.shortest_path_length(L, 'JK')
```

The response to the preceding line of code is shown, as follows:

```
{'Aaron': 3,
 'Ashish': 1,
 'Athulya': 1,
 'Derrick': 2,
 'Enoch': 2,
 'Eric': 1,
 'JK': 0,
 'Jeremiah': 2,
 'Obama': 3,
 'Peter': 1,
 'Vijay': 2}
```

The preceding dictionary returned by the shortest_path_length() method lets us know the shortest path length for JK to everyone else in the network. Of course, the shortest path length to JK himself is 0.

Now, the preceding shortest path algorithm is operating on edges with unit weight: that is, there is an assumption that all of the connections are equally likely to connect the source to the target. And LinkedIn probably doesn't weight your connections; it doesn't need to. But other types of networks, such as a road network, isn't unweighted networks. They are, in fact, weighted directional networks.

Shortest path algorithms also work on such graphs and try to optimize for edge weight. In order words, they try to provide a solution for traversing from one node to another target node by accumulating the minimum cost, in terms of edge weight. This sounds very intuitive. An experienced driver in a city doesn't need any navigation app to take the shortest path between any two points: they intuitively do that. But it is highly improbable that the driver is able to factor in current traffic or road accidents and dynamically change their route or drive in the shortest time possible to a location they've never been to. For novice drivers and tourists, they are absolutely dependent on a navigation app to solve their shortest route problem. So, let's solve it for them. There are a lot of shortest paths algorithms, but let's review the basics of one of the most popular shortest path algorithm before jumping into the code.

Dijkstra's algorithm

If you find *Dijkstra* to be a tongue twister, you are like everyone else in the world minus the Dutch. But this guy with a tongue-twister name is credited with introducing one of the most efficient algorithms for solving routing problems, thereby giving the algorithm his name as well as tongue-sprain (if such a thing exists) for generations of computer scientists to come. Let's understand how Dijkstra's algorithm works using a simple network represented as follows:

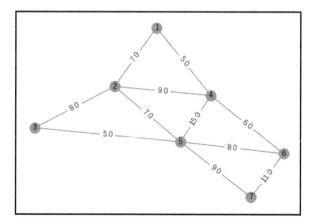

A connected undirected graph

Let's represent this graph as a list of tuples, as shown in the following code. The first two elements in each tuple represent the nodes being connected and the third value represents the weight of the edge between the corresponding nodes:

```
eList = [(1, 2, 7.0), (1, 4, 5.0),
(2, 3, 8.0), (2, 4, 9.0), (2, 5, 7.0),
 (3, 5, 5.0), (4, 5, 15.0), (4, 6, 6.0),
 (5, 6, 8.0), (5, 7, 9.0), (6, 7, 11.0)]
```

The `networkx` library can ingest this list of tuples and construct a graph from it using the `add_weighted_edges_from()` method:

```
R = nx.Graph()
R.add_weighted_edges_from(eList)
```

Calculating Dijkstra's shortest path

Let's say our goal is to reach node 7 from node 1 by accumulating the minimum cost. Let's ask the graph right away what's the shortest path from node 1 to 7:

```
nx.dijkstra_path(R, 1, 7)
```

We get the following response:

```
[1, 4, 6, 7]
```

This means that the shortest path to reach node 7 from node 1 is through nodes 4 and 6 respectively. The traversal one has to make is as follows: edge 1-4, then edge 4-6 and, finally, edge 6-7.

Calculating Dijkstra shortest path length

If the graph were an unweighted graph, the shortest path length would have been three, since only three *hops* were needed to reach node 7 from node 1. Since it is a weighted graph, to calculate the accumulated cost, we need to add up the weights of the edges 1–4, 4–6, and 6–7, which is 5.0, 6.0, and 11.0 respectively. Hence, the shortest path length is 22.0. Let's confirm this with the `dijkstra_path_length()` method, as follows:

```
nx.dijkstra_path_length(R, 1, 7)
```

The response is, as expected, 22.0.

The arguments for the preceding function are the graph object, the source node, and the target node respectively. When in doubt, always enter the method name prefixed by the `?` sign in a Colab (or Jupyter) Notebook cell:

```
?nx.dijkstra_path_length()
```

Once we type and execute this code, we should see something similar open up in the Notebook:

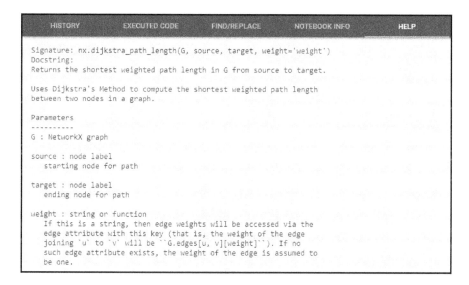

In `docstring`, you might have noticed that there's an optional argument, `weight`. Since the default value is `weight` itself, we didn't have to mention this explicitly.

If we need to verify the data attributes present in the edges, we can just reckon the edge data from the graph object, as follows:

```
R.edges(1, data=True)
```

The following response confirms that we have data for our `edges` with the `weight` attribute name, at least for edges incident to node `1`:

```
EdgeDataView([(1, 2, {'weight': 7.0}), (1, 4, {'weight': 5.0})])
```

 Remember, the first argument in the preceding (`R.edges()`) method is also optional. We provided the node ID as `1` so that the preceding response is truncated. Try to look at the response without the optional first and second arguments as well.

Calculating single source Dijkstra path length

Let's try something even more interesting. Let's try to calculate the Dijkstra's shortest path length to each and every node from the source node, that is, node 1. We can accomplish this using the following method: `nx.single_source_dijkstra_path_length()`.

The signature of the method looks like this:

```
nx.single_source_dijkstra_path_length(G, source, cutoff=None,
weight='weight')
```

Of course, feel free to look up the entire `docstring` by prepending the method name with a single question (?), or if you want to venture into the source code of the method, prepend a double-question mark (??).

The method signature suggests that the graph object and the source node are required parameters and the two other parameters are optional. By now, we know what the weight parameter is. The `cutoff` parameter accepts a `cutoff` value for the accumulated weights. We'll try to understand this once we execute the method without the optional parameters first:

```
nx.single_source_dijkstra_path_length(R, 1)
```

The response to the preceding method is as follows:

```
{1: 0, 4: 5.0, 2: 7.0, 6: 11.0, 5: 14.0, 3: 15.0, 7: 22.0}
```

In the response, the keys of the dictionary represent the node IDs and the values of the dictionary items represent the shortest path length to the corresponding node from the source node. From our earlier queries, we know very well that the shortest distance length to node 7 from node 1 is 22.0 and the distance to node 1 itself is 0, which can be observed in the preceding response.

Let's try to get the shortest path lengths to the entire network from a different source node, say, node 5, and look at the response:

```
nx.single_source_dijkstra_path_length(R, 5)
```

The following response provides us with some naive facts, such as the fact that the shortest path distance to itself is 0, and some interesting insights such as the furthest nodes from node 5 are node 1 and node 4; the cost to reach both of these nodes is 14.0:

```
{5: 0, 3: 5.0, 2: 7.0, 6: 8.0, 7: 9.0, 1: 14.0, 4: 14.0}
```

Let's assume that that the accumulated weight values in the preceding response refers to the distance, in miles, it takes to travel from location 5 to all possible neighboring locations. Say we are evaluating all possible locations neighboring location 5, but we have a hard limit of **10** miles that a location in the network can be away from location 5. How will the preceding method call change, when we need to introduce such a criterion? It's here that the cutoff parameter comes into play. Plug in the value of 10.0 to the cutoff parameter in the preceding code. The scenario is illustrated as follows:

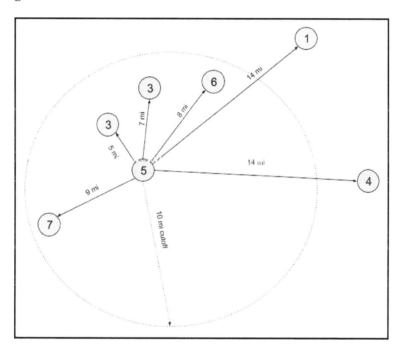

Assuming graph weights as distance in miles and introducing a cut-off

The code implementation for the scenario translates as shown as follows:

```
nx.single_source_dijkstra_path_length(R, 5, cutoff=10)
```

Executing the preceding line of code provides us with the following output:

```
{5: 0, 3: 5.0, 2: 7.0, 6: 8.0, 7: 9.0}
```

As you would expect, node 1 and node 4 are excluded in the preceding response. This is quite useful while querying large graphs and we will be using this method to create something known as an isochrone a little later.

Turning a simple DataFrame into graphs

We dealt with constructing a graph by defining the individual edges and an edge list manually. This can be cumbersome for large graphs. But we can import these from other sources, such as edge list files and noticeably from a pandas DataFrame. The DataFrame just needs to have a source column, a target column, and an optional weight column.

Let's start with a simple pandas DataFrame and then deal with larger data frames very soon:

```
import pandas as pd
edges = pd.DataFrame({ 'source': [0, 1, 2],
                       'target': [2, 2, 3],
                       'dist': [3, 4, 5],
                       'time' :[10, 12, 15]
                     })
edges
```

When it is displayed, the pandas DataFrame looks like this. The source and target columns probably represent the edges and the directionality of the edges. So, this should definitely be converted into a digraph (nx.DiGraph()). There are two other attribute fields, dist and time, both of which could potentially act as a weight of the edges:

	dist	source	target	time
0	3	0	2	10
1	4	1	2	12
2	5	2	3	15

A simple pandas DataFrame that could be converted to a graph

A method called nx.from_pandas_edgelist() can be used to convert a pandas DataFrame into a graph:

```
G = nx.from_pandas_edgelist(edges, edge_attr=True, create_using=
nx.DiGraph())
```

In the preceding code, we are passing edges to the from_pandas_edgelist() method and setting the edge_attr parameter to be True. We are also indicating that a digraph should be created from the DataFrame through the create_using parameter. That's it: we have converted a DataFrame into a networkx graph, upon which we can perform all kinds of graph operations.

The following lines of code can be used to visualize a graph:

```
pos=nx.spring_layout(G)
nx.draw_networkx(G, pos, edge_color='b', )
edge_labels = nx.get_edge_attributes(G,'time')
nx.draw_networkx_edge_labels(G, pos, font_size=12, edge_labels=edge_labels)
plt.axis('off')
```

In the following visualization, the time attribute of the edges is displayed:

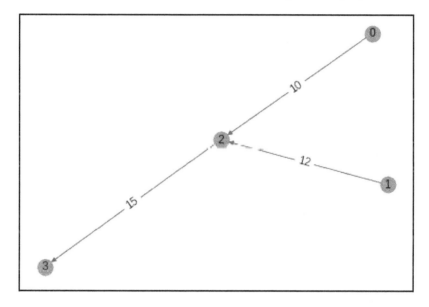

Visualizing the previous graph by considering time as the weight of the graph

We have gained a foundational knowledge about graph data structure, its representation using the `networkx` library, and about a few popular analyses performed using graph data structure. In the next section, we will deal with leveraging this data structure to perform analyses on spatial data.

Building a graph based on a road network

In the previous sections, we learned about the fundamentals of the graph and many of the operations that we can perform on a graph using sample data that we defined. In this section, we will discuss how to import real-world data such as a road network and translate it into a graph and perform analyses on that. For this purpose, we will be importing sample **Open Street Maps (OSM)** road data that we have provided with this chapter.

Open Street Maps data

OSM provides a rich set of open source and crowdsourced geographic data of many parts of the world. In particular, we are interested in the road data they provide. For the US, they are baselined on census TIGER road shapefiles and benefit from continuous updates by the active OSM community. OSM road data provide critical information such as the following:

- Road geometry
- Road direction
- Road speed limit
- Traffic signs and impediments (pedestrian crossings and so on)

These pieces of information will help us to build robust road networks and perform routing analyses on that. We will read the sampled OSM data from the data repository and convert it into a graph.

Exploring the road data

For reading the shapefiles, we will be using the GeoPandas library and Shapely geometry modules. If GeoPandas is not installed, install it using `pip` or any similar package manager. The `pip` installation code is shown as follows:

```
!pip install geopandas
```

To execute the code further, please download the `BayArea4` shapefiles from our data repository into the folder where your Jupyter Notebook resides.

Let's now import the rest of the libraries required to create our graph:

```
import geopandas as gpd
import matplotlib.pyplot as plt
from shapely.geometry import LineString
import pandas as pd
```

If you have uploaded the `BayAreaRoads` shapefile into your folder as we instructed, you can reuse the following lines of code; otherwise, you might need to modify your code as per the location of your data upload:

```
roads_df = gpd.read_file('Roads_BayArea4.shp')
```

Like we have discussed in previous chapters, the previous line of code loads the shapefile as a GeoDataFrame, which is nothing but a DataFrame with a geometry column. Let's try to display the GeoDataFrame like we do for a normal pandas DataFrame for preliminary data exploration:

```
roads_df.tail()
```

The following snapshot shows a DataFrame view of our roads DataFrame. All of the columns of the DataFrame except the geometry column are readable and provide us with much more information about the features they represent. From a preliminary observation, we recognize that each row represents a road segment and other columns such as name and fclass provide the name and type of the road. For more information on the field and format used, you should refer to Geofabrik's documentation provided in the PDF document here https://download.geofabrik.de/osm-data-in-gis-formats-free.pdf:

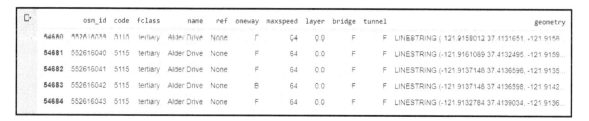

DataFrame view of roads GeoDataFrame

The maxspeed column is of particular interest since it defines the maximum speed in which someone can traverse that road segment as permitted by law. If we need to create a time-based routing, we need to leverage this field. But before doing that, we need to make sure that the field provides us with complete information. To do that, let's execute the following code, which provides us with the sum of rows with null values in each column:

```
roads_df.isnull().sum(axis=0)
```

The response to the preceding line of code is as follows:

```
osm_id 0
code 0
fclass 0
name 8975
ref 47125
oneway 0
maxspeed 0
layer 0
bridge 0
tunnel 0
```

```
geometry 0
dtype: int64
```

A first look at the preceding response looks promising. None of the rows in the `maxspeed` column have null values. However, it might be possible that most of the null values for the `maxspeed` column are substituted by 0, which is an unacceptable value if we were to use the `maxspeed` column to derive the time taken to traverse a road segment. Let's confirm that this isn't the case:

```
len(roads_df[roads_df["maxspeed"] == 0])
```

Unfortunately, we get a response that is far greater than zero:

```
34952
```

35,000 out of 54,000 records have 0 as a value for `maxspeed`. So, we need to find a proxy to calculate the values for `maxspeed`. This will be discussed in detail in the following sections. For now, let's try to visualize the roads `GeoDataframe` as a simple plot using the following lines of code:

```
%matplotlib inline
roads_df.plot(figsize=(10,10))
```

The simple plot of the roads' `GeoDataFrame` shown here reveals to us the line geometry of the roads and the network it forms:

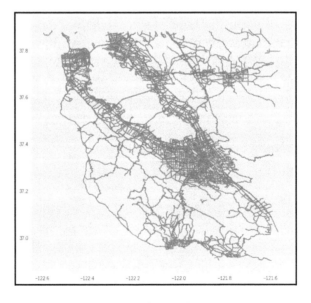

Graph view of roads DataFrame

Creating a graph from a DataFrame

The following steps are required to create a graph from our `DataFrame`:

1. Reading and exploding the geometry
2. Calculating the distance of each edge in meters
3. Finding a proxy for zero values of `maxspeed`
4. Accounting for directionality
5. Calculating drive time in seconds for each edge

Reading and exploding the geometry

The `geometry` column holds the key to constructing the edges required for our road network graph. Let's read just a random row for and play around with the `geometry` column and try to make sense of it:

```
frow = roads_df.iloc[1000]
print("{frow}\n\n")
print("Coords : {frow['geometry'].xy}\n\n")
print("Length {frow['geometry'].length}")
```

We picked up the 1,001st row and tried to print out the information in the row and the geometric information such as the coordinates in the geometry `LineString` and the length of the `LineString`. The length of the `LineString` is in decimal degrees and doesn't make much sense and is not useful to us, but the coordinates of the `LineString` are. The coordinates are available as a tuple of NumPy arrays, with the first item in the tuple being the NumPy array of longitudes and the second item being the array of latitudes:

```
osm_id                    6402644
code                      5115
fclass                    tertiary
name                      Almaden Boulevard
ref                       None
oneway                    B
maxspeed                  40
layer                     0
bridge                    F
tunnel                    F
geometry LINESTRING (-122.0593666 37.6038845, -122.0581...
Name: 1000, dtype: objectName: Almaden Boulevard

Coords : (array('d', [-122.0593666, -122.0581518, -122.0572934,
-122.0571144, -122.0569186, -122.0567657, -122.0566584, -122.0566088,
```

```
-122.0567335, -122.056928, -122.057396, -122.0574698, -122.0574577,
-122.0573591, -122.0572354, -122.0566423, -122.0561337, -122.0555037,
-122.0553468, -122.0551805, -122.0550518, -122.0549095, -122.0548292,
-122.0548359, -122.0550356, -122.0550716]), array('d', [37.6038845,
37.6036281, 37.6034445, 37.6033889, 37.6032577, 37.6031047, 37.6029134,
37.6026462, 37.6022037, 37.6015077, 37.5998034, 37.5995144, 37.5991935,
37.5990012, 37.5988448, 37.5984593, 37.5981822, 37.5979971, 37.5979355,
37.5978271, 37.5977187, 37.5975808, 37.5973064, 37.5970493, 37.5965537,
37.5964397]))

Length 0.01110826653688575
```

How do these coordinates look in the real world? Well, wonder no more; a simple matplot plotting will answer that:

```
x, y = frow['geometry'].xy
fig = plt.figure(1, figsize=(5,5), dpi=90)
ax = fig.add_subplot(111)
ax.plot(x, y)
```

The output will be as follows:

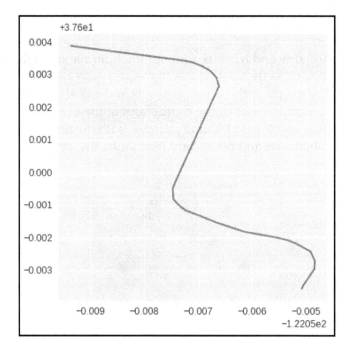

A segment of the Almaden Boulevard plotted using matplotlib

Twenty-six coordinates were to create this sinuous road segment. This means that 25 graph edges be created from the 26 coordinates.

> The coordinate values (from the `frow['geometry'].xy` object) are in this format:
> $(array('d', [X_0, X_1, X_2, ...X_{25}]), array('d', [Y_0, Y_1, Y_2, ...Y_{25}]))$
>
> However, we need to consolidate the coordinates in this format :
> $[[X_0, Y_0], [X_1, Y_1], [X_2, Y_2],...,[X_{25}, Y_{25}]]$

> The preceding format represents a list of nodes in each road segment. We need to be able to create the graph edges from the node list so that the edges look like this:
>
> *Edge 1 : $[[X_0, Y_0], [X_1, Y_1]]$*
> *Edge 2 : $[[X_1, Y_1], [X_2, Y_2]]$*
> *Edge 3 : $[[X_2, Y_2], [X_3, Y_3]]$*
> *to*
> *Edge 25 : $[[X_{24}, Y_{24}], [X_{25}, Y_{25}]]$*

The following lines of codes accomplish these requirements.:

```
def split_data_frame_list(df, target_column, col_name, output_type=list):
    '''
    Accepts a column with multiple types and splits list variables to
several rows.
    df: dataframe to split
    target_column: the column containing the values to split
    col_name : name of the split column
    output_type: type of column to be split
    returns: a dataframe with each entry for the target column separated,
with each element moved into a new row.
    The values in the other columns are duplicated across the newly divided
rows.
    '''
    row_accumulator = []

    def split_list_to_rows(row):
        split_row = row[target_column]
        if isinstance(split_row, list):
            for s in split_row:
                new_row = row.to_dict()
                new_row[col_name] = output_type(s)
                row_accumulator.append(new_row)
        else:
```

```
            new_row = row.to_dict()
            new_row[col_name] = output_type(split_row)
            row_accumulator.append(new_row)

    df.apply(split_list_to_rows, axis=1)
    new_df = pd.DataFrame.from_records(row_accumulator,
columns=[df.columns])
    return new_df
```

The preceding function explodes the road segments into individual rows, so that we have one edge per row. After defining the preceding function, we can execute the code provided as follows. This will create a new DataFrame: edge_df. edge_df will contain a geometry column that contains the edge coordinates. Two separate columns, source and target, should also be created. The source column will contain the first coordinate of the edge, and the target column will contain the second coordinate of the edge:

```
roads_df["geometry_arr"] = roads_df["geometry"].apply(lambda geom :
list(zip(*geom.xy)))
roads_df["edge_list"] =  roads_df["geometry_arr"].apply(lambda coords :
[[coords[i], coords[i+1]] for i in range(len(coords) - 1)])

edge_df = split_data_frame_list(roads_df, "edge_list", "geometry")
edge_df["source"] = edge_df["geometry"].apply(lambda edge: edge[0])
edge_df["target"] = edge_df["geometry"].apply(lambda edge: edge[1])
edge_df = edge_df.drop('geometry', axis = 1)
edge_df = edge_df.drop('geometry_arr', axis = 1)
roads_df = roads_df.drop(["geometry_arr", "edge_list"], axis = 1)
edge_df.head()
```

The newly created edge_df should look like this:

	osm_id	code	fclass	name	ref	oneway	maxspeed	layer	bridge	tunnel	source	target
0	4304424	5134	secondary_link	None	None	F	0	1.0	T	F	(-122.0470346, 37.3867598)	(-122.0469872, 37.3867158)
1	4304424	5134	secondary_link	None	None	F	0	1.0	T	F	(-122.0469872, 37.3867158)	(-122.04691, 37.3866479)
2	4304424	5134	secondary_link	None	None	F	0	1.0	T	F	(-122.04691, 37.3866479)	(-122.046828, 37.3865875)
3	4304424	5134	secondary_link	None	None	F	0	1.0	T	F	(-122.046828, 37.3865875)	(-122.0467464, 37.3865318)
4	4304424	5134	secondary_link	None	None	F	0	1.0	T	F	(-122.0467464, 37.3865318)	(-122.0466566, 37.3864817)

Newly created edge_df

Calculating the distance of edges

We can use the `haversine` function discussed in earlier chapters to calculate the distance for each edge. The following lines of code will add a `dist` column to `edge_df` and populate it with the length of the edge in meters:

```python
from math import radians, cos, sin, asin, sqrt, atan2

def haversine(loc1, loc2):
    """
    Calculate the great circle distance between two points
    on the earth (specified in decimal degrees)
    """
    lon1, lat1 = loc1
    lon2, lat2 = loc2
    # convert decimal degrees to radians
    lon1, lat1, lon2, lat2 = map(radians, [lon1, lat1, lon2, lat2])

    # haversine formula
    dlon = lon2 - lon1
    dlat = lat2 - lat1
    a = sin(dlat/2)**2 + cos(lat1) * cos(lat2) * sin(dlon/2)**2
    c = 2 * asin(sqrt(a))
    r = 6378100 # Radius of earth in meters. Use 3956 for miles
    return c * r


edge_df["dist"] = edge_df.apply(lambda row: haversine(row["source"],
row["target"]), axis = 1)
```

Finding a proxy for maximum speed

Our job, at this point, is to find a proxy for maximum speed values when the value is not present (or populated by zero). The `fclass` column provides the best opportunity to impute this data. The data we are working with corresponds to the state of California. The following speed lookup for the `fclass` type holds good enough for California:

```python
speed_limits_in_mph = {
    "living_street"   : 10,
    "motorway"        : 65,
    "motorway_link"   : 40,
    "primary"         : 55,
    "primary_link"    : 30,
    "residential"     : 25,
    "secondary"       : 35,
    "secondary_link"  : 25,
```

```
    "service"            : 15,
    "tertiary"           : 35,
    "tertiary_link"      : 25,
    "trunk"              : 55,
    "trunk_link"         : 30,
    "unclassified"       : 55,
    "unknown"            : 40
}
```

Update the `maxspeed` column in `edge_df` with the preceding lookup dictionary. Remember, the `maxspeed` column is in kilometers per hour and the lookup table speeds are provided in miles per hour. So, while updating `maxspeed`, a multiplicative factor of `1.609` is added to the lookup table speeds (since 1 mile = 1.609 km):

```
edge_df["maxspeed"] = edge_df.apply(lambda row:
speed_limits_in_mph[row["fclass"]] * 1.609 if row["maxspeed"] == 0 else
row["maxspeed"], axis = 1)
```

Accounting for directionality

The directionality of the edges is defined in the `oneway` column. The definition for the column is provided in the data dictionary provided by the Geofabrik download site.

The following indicates different notations used in the code:
oneway: Is this a one-way road?
F: Only driving in the direction of the `LineString` is allowed.
T: Only the opposite direction is allowed.
B: Both directions are OK.

For edges with `oneway == 'F'`, no changes are required. For edges with `oneway == 'T'`, the current edges have to flip, that is, the `source` and `target` values must be swapped. For edges with `oneway == 'B'`, additional edges replicating the edge need to be added and then swapped. The following lines of code will achieve the preceding objective:

```
pd.options.mode.chained_assignment = None
nt = edge_df.query("oneway != 'T'", inplace = False)
b = edge_df.query("oneway == 'B'", inplace = False)

src_tmp = b["source"]
b["source"] = b["target"]
b["target"] = src_tmp

t = edge_df.query("oneway == 'T'" , inplace = False)
src_tmp = t["source"]
t["source"] = t["target"]
```

```
t["target"] = src_tmppd.options.mode.chained_assignment = None
nt = edge_df.query("oneway != 'T'", inplace = False)
b = edge_df.query("oneway == 'B'", inplace = False)

src_tmp = b["source"]
b["source"] = b["target"]
b["target"] = src_tmp

t = edge_df.query("oneway == 'T'" , inplace = False)
src_tmp = t["source"]
t["source"] = t["target"]
t["target"] = src_tmp

edge_df = pd.concat([nt, b, t], ignore_index=True)
```

Calculating drivetime

Drivetime for the edges can be calculated using the middle school *Time* and *Distance* calculations. The following formulae and conversion will help our time calculations:

$$Time = Distance / Speed$$
$$Time \ (in \ seconds) = (Distance \ (in \ meters) / Speed \ (in \ kmph) \ ) * (18/5)$$

Let's implement this formula with the following line of code:

```
edge_df["time"] = (edge_df["dist"]/edge_df["maxspeed"]) * (18/5)
```

We have hence effectively added a `time` column. No more processing is needed for the `DataFrame` at this point. We are ready to convert this into a graph:

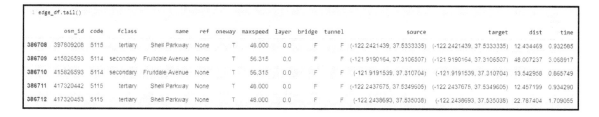

	osm_id	code	fclass	name	ref	oneway	maxspeed	layer	bridge	tunnel	source	target	dist	time
386708	397809208	5115	tertiary	Shell Parkway	None	T	48.000	0.0	F	F	(-122.2421439, 37.5333335)	(-122.2421439, 37.5333335)	12.434469	0.932585
386709	415826593	5114	secondary	Fruitdale Avenue	None	T	56.315	0.0	F	F	(-121.9190164, 37.3106507)	(-121.9190164, 37.3106507)	48.007237	3.068917
386710	415826593	5114	secondary	Fruitdale Avenue	None	T	56.315	0.0	F	F	(-121.9191539, 37.310704)	(-121.9191539, 37.310704)	13.542958	0.865749
386711	417320442	5115	tertiary	Shell Parkway	None	T	48.000	0.0	F	F	(-122.2437675, 37.5349605)	(-122.2437675, 37.5349605)	12.457199	0.934290
386712	417320453	5115	tertiary	Shell Parkway	None	T	48.000	0.0	F	F	(-122.2438693, 37.535038)	(-122.2438693, 37.535038)	22.787404	1.709055

Building the graph

We have already discussed how to convert a pandas `DataFrame` into a graph using the `networkx` method known as `from_pandas_edgelist()`. Since our `edge_df` already contains the `source` and `target` columns, we need not mention it explicitly in the method. To keep the footprint of the graph light, let's only include the most essential data from the `DataFrame` as `edge` attributes. Here, we are adding `osm_id`, `name`, and `time` as edge attributes:

```
G = nx.from_pandas_edgelist (df = edge_df, edge_attr = ["osm_id", "name",
"time"], create_using  = nx.DiGraph())
```

With that preceding line of code, we have built a graph from a road network `GeoDataframe`. In the next section, we will be doing all of the exciting stuff such as performing analyses with this graph.

Shortest path analyses on the road network graph

In this section, we will be performing some of the analyses that we did earlier on the new graph, such as the following:

- Dijkstra's shortest path analysis
- Dijkstra's shortest path cost
- Single-source Dijkstra's path length

Dijkstra's shortest path analysis

We have waited so long to perform the shortest path analysis on our graph. So, without further ado, let's run the shortest path algorithm on our graph, optimizing for time. To run the shortest path algorithm, we need to know the source node ID and target node ID. Here, our node IDs are coordinates. Right now, let's pick the node IDs randomly:

```
nodelist = list(G.nodes())
import random
random.seed(0)
```

The following lines of code select two nodes at random and assign them as the source and the target node, respectively. This is used as input to the `dijkstra_path()` function. As we have seen earlier, this function returns a list of node IDs in order, joining the ones that will make up the shortest path from the source node to the target node. Since the returned node IDs are already coordinates, we can create a Shapely `LineString` from it and plot it on the roads `GeoDataFrame`:

```
# Selecting random source and target
src, target = nodelist[random.randint(0, len(nodelist)-1)],
nodelist[random.randint(0, len(nodelist)-1)]

#Calculating Shortest path
sp = nx.dijkstra_path(G, src, target, weight='time')

#Creating Shortest Path geometry with the result
spd = gpd.GeoDataFrame({"geometry" : gpd.GeoSeries(LineString([(item) for
item in sp]))})

#Plot the geometry
ax = roads_df.plot(color='gray', figsize = (12, 12), linewidth = 0.7, alpha
= 0.5)
spd.plot(ax=ax, color='black', linewidth=2)

ax.text(src[0], src[1], s = 'Source', fontsize=12 ,color='r')
ax.text(target[0], target[1], s = 'Target', fontsize=12, color='r')
ax.scatter([src[0], target[0]],[src[1], target[1]], marker = '^', s = 100)
#Markers

ax.axis('equal')
ax.axis('off')
```

This code generates a plot as follows. The dark line represents the shortest path between a source and target node in the road graph (represented by gray lines):

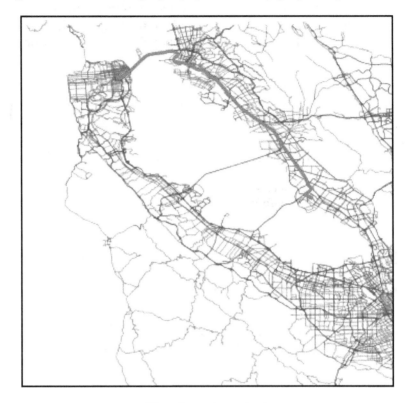

Dijkstra's shortest path on a road graph

Dijkstra's shortest path cost

To get the time cost of travel from the source node to the target node, just use the following one-liner:

```
nx.dijkstra_path_length(G, src, target, weight='time')//60
```

The response we get is `36.0` (in minutes) since we have made an integer division by `60`.

Single source Dijkstra's shortest path cost

When we need to know the cost to reach all possible nodes from the source node, with a maximum accumulated cost of 30 minutes (1800 seconds), we can use the `single_source_dijkstra_path_length()` method, as follows:

```
spl = nx.single_source_dijkstra_path_length(G, src, cutoff = 1800,
weight="time")
dict([spl.popitem() for i in range(0, 10)])
```

The response of the method, stored in the `spl` variable, is a Python dictionary with the node IDs as the keys of the dictionary item and corresponding time cost (in seconds) to reach the node from the defined source node as dictionary values. The last ten items of the dictionary are displayed, as follows:

```
{
(-122.1837371, 37.4844172): 1799.9573436643684,
(-122.1168455, 37.4255555): 1799.976275047492,
(-122.0215445, 37.545765): 1799.9808136253362,
(-122.0044161, 37.5395224): 1799.6705228569708,
(-121.9572673, 37.5326912): 1799.7428166080235,
(-121.9550996, 37.524291): 1799.6916229635788,
(-121.9525275, 37.5363582): 1799.870403332123,
(-121.9460679, 37.5279106): 1799.7422486517696,
(-121.9309729, 37.5279041): 1799.6927213642161,
(-121.6509853, 37.1529573): 1799.9385337063288
}
```

Concept of isochrones

Shortest paths are a navigational solution that provides the shortest paths and the shortest time (or distance) from a source to a target node. What if we need a solution with all possible areas that we can reach from a source node within a given time or distance cut-off? This is very important in commute-based searches, which can answer questions such as the following:

- What are the nearest coffee shops within 15 minutes by car?
- What are the open home listings that are 30 minutes from my office?
- Can any fire truck stationed at its base location reach my house within 10 minutes?
- What are the areas that are under-represented by public transit agencies?

The idea we are looking at is known as an isochrone. Etymologically, it means *same time* (*iso* = same and *chrono* = time). More specifically, it means the areas that you can reach within the same time from a source location by a particular mode of transport under normal circumstances:

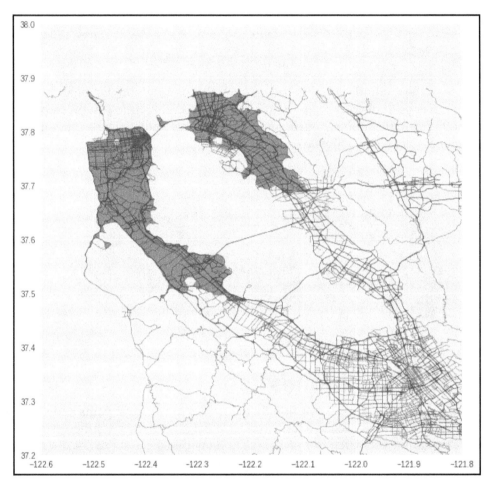

Isochrone

The preceding graph shows the isochrone as a red, shaded area. The isochrone represents all possible locations that we can drive to from the yellow dot shown, as per our graph calculations.

Constructing an isochrone

As you might have guessed, we can use the response from the `single_source_dijkstra_path_length()` function to create an isochrone. But once we execute this method, we might have to further construct a polygon from the returned node locations. There are different methods to create a polygon given a mesh of points, but we will be discussing an approach known as the concave hull. Concave hulls are created by repeated triangulation of available points in the most efficient way. The implementation of this triangulation is provided by a `scipy.spatial` module known as `Delaunay`. The process of creating an isochrone polygon is as follows:

1. Create Delauney triangles with the location of the result nodes, as shown here:

   ```
   # Compute Delaunay triangles
   tri = Delaunay(coords)
   ```

2. Loop through each triangle and calculate the circumradius of the triangle
3. If the circumradius of the triangle is less than a preset value (say, 1/150), add the vertices of the triangle to a set
4. Construct a series of polygons from the set of vertices
5. Merge the polygons

The code implementation of this logic can be found in the code repository under a function named `generate_alpha_poly()`. Once we have defined the alpha polygon generator function, we can use the response from our `single_source_dijkstra_path_length()` method, stored in the `spl` variable. Since we are using the `cutoff` parameter, we just need the keys of the `spl` dictionary. For a 30-minute isochrone, we will receive a lot of nodes as coordinates; hence, it is OK to round off the decimal degrees to 3 digits in the node coordinates:

```
pts = np.array(list(spl.keys()))
pts = np.round(pts, 3)
pts = np.unique(pts,axis = 0)
poly = generate_alpha_poly(pts, 150)
```

When this polygon is plotted, it looks exactly like a plot that was shown at the beginning of this section: the diagram titled *Isochrone*.

Summary

We started off with creating a simple graph using the `networkx` library and ended up by creating isochrones from real-world road network data. We also explored the various functionalities offered by the `networkx` library to solve graph problems such as shortest path and shortest path length. We also went into depth to understand the geometric and data transformations required to translate a `GeoDataFrame` into a graph data structure. The best part of the entire chapter was that we were able to do all of these using just open source data and tools. Just by leveraging the skills we gained so far, we were able to create many insights that are invaluable to a wide range of industries.

In the next chapter, we will transition into building location recommender systems using the concepts we have dealt with so far, as well as integrating it with state-of-the-art machine learning and deep learning techniques used for recommendation systems. We will be building simple route recommenders and travel time predictors, as well as venue recommenders. We will also deal with how to evaluate such systems in the context of geospatial data.

7
Getting Location Recommender Systems

Recommendation systems are primarily used to predict the preference or rating of a user for an item. They are widely used in many commercial applications, including product and service recommendations, as well as content and friendship recommendations in social media. However, recommendation systems are not only used for products on Amazon or movies on Netflix but also locations. Location-based recommenders incorporate the location of users to provide relevant and precise recommendations. These can be a point of interest recommendations, such as restaurants, events in nearby locations, or posts and local trends in social media. In this chapter, we will cover different recommender systems, including collaborative filtering methods and location-based recommendation methods. We will take an example of a restaurant recommender system application in this chapter, using a restaurant and consumer dataset from UCI, Machine Learning Repository.

 Citation: *Blanca Vargas-Govea, Juan Gabriel GonzÃ¡lez-Serna, Rafael Ponce-MedellÃn. Effects of relevant contextual features in the performance of a restaurant recommender system. In RecSysâ€™11: Workshop on Context Aware Recommender Systems (CARS-2011), Chicago, IL, USA, October 23, 2011.*

The topics covered in this chapter include the following:

- Exploratory data analysis
- Collaborative filtering recommenders
- Location-based recommendation systems

Exploratory data analysis

Let's start reading the data. We will be using two files: one CSV with ratings and another GeoJSON file with restaurants and their locations. Let's first read the ratings of the CSV file.

Rating data

This file contains the final rating of restaurants. It has `userID` and `placeID`, which we can merge with the GeoJSON datasets of restaurants and rating columns. Let's read the data in pandas and look at the first five rows:

```
ratings = pd.read_csv('RCdata/rating_final.csv')
ratings.head()
```

The table looks like this, with a rating of each user for some restaurants:

	userID	placeID	rating	food_rating	service_rating
0	U1077	135085	2	2	2
1	U1077	135038	2	2	1
2	U1077	132825	2	2	2
3	U1077	135060	1	2	2
4	U1068	135104	1	1	2

User ratings

We have 1,161 rating rows and if we look at the first five rows of the `rating` column, the first three rows under the `rating` column have a 2 point rating, while the last two rows have a 1 point rating. Let's get the mean of the rating column by using the pandas `.mean()` function:

```
print(ratings['rating'].mean())
```

The output the preceding code shows that the average rating of the entire rating dataset is 1.20. Let's print out how many unique `userID` we have, as well as unique `placeID`:

```
print("There are {} unique userID in the
dataset".format(ratings['userID'].nunique()))
print("There are {} unique placeID in the
dataset".format(ratings['placeID'].nunique()))
```

The output of the preceding code prints out the following and we can see that we have 138 unique `userID` and 130 unique `placeID`:

```
There are 138 unique userID in the dataset
There are 130 unique placeID in the dataset
```

We can also go further and look at `countplot` to get the distribution of the ratings. We will use `seaborn` for our data visualization in this chapter:

```
fig, ax = plt.subplots(figsize=(12,10))
sns.countplot(ratings['rating'], ax=ax)
plt.show()
```

The output of the preceding code is `countplot`, where the total of each rating number is calculated. This plot shows that a rating range of **0** to **2** is available for this dataset and most restaurants have a rating of **2**, while around more than 250 restaurants have **0** ratings:

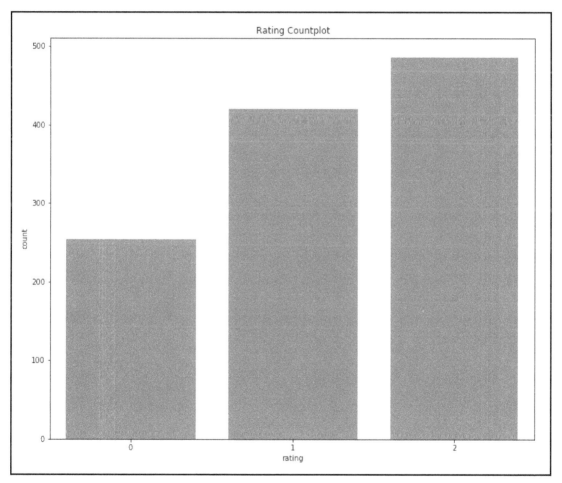

Rating countplot

In the next section, we will also explore the restaurant dataset.

Restaurants data

The restaurant dataset is in GeoJSON format and therefore we do not need to convert it into `GeoDataFrame` but rather can read it directly with GeoPandas. Let's do that. The data comes with 22 columns and, therefore, it will be difficult to read the first five rows in the way that we normally do in this book. We will only read the first two rows and transpose the output to fit it into the screen:

```
# Read the data as GeoDataFrame
geoplaces = gpd.read_file('RCdata/geoplaces.geojson')
geoplaces.head(2).T
```

Here is the output of the first two rows transposed. The columns are now displayed as rows and vice versa. This data comes in 130 rows, matching `placeID` in the `ratings` datasets:

	0	1
placeID	134999	132825
latitude	18.9154	22.1474
longitude	-99.1849	-100.983
the_geom_meter	0101000020957F000088568DE356715AC138C0A525FC46...	0101000020957F00001AD016568C4858C1243261274BA5...
name	Kiku Cuernavaca	puesto de tacos
address	Revolucion	esquina santos degollado y leon guzman
city	Cuernavaca	s.l.p.
state	Morelos	s.l.p.
country	Mexico	mexico
fax	?	?
zip	?	78280
alcohol	No_Alcohol_Served	No_Alcohol_Served
smoking_area	none	none
dress_code	informal	informal
accessibility	no_accessibility	completely
price	medium	low
url	kikucuernavaca.com.mx	?
Rambience	familiar	familiar
franchise	f	f
area	closed	open
other_services	none	none
geometry	POINT (-99.184871 18.915421)	POINT (-100.983092 22.1473922)

Restaurant dataset

Since we have a `geometry` column and read the data with GeoPandas, we can plot this data as a map. Let's do a clustered map with `folium`:

```
lons = geoplaces['longitude']
lats = geoplaces['latitude']
m = folium.Map(
 location = [np.mean(geoplaces.latitude), np.mean(geoplaces.longitude)],
 tiles= 'CartoDB dark_matter',
 zoom_start=6
 )
FastMarkerCluster(data=list(zip(lats, lons))).add_to(m)
folium.LayerControl().add_to(m)
m
```

The output of the clustered map is as follows. As you can see, this dataset is in Mexico and, as the clusters show, it has three distinct places within Mexico:

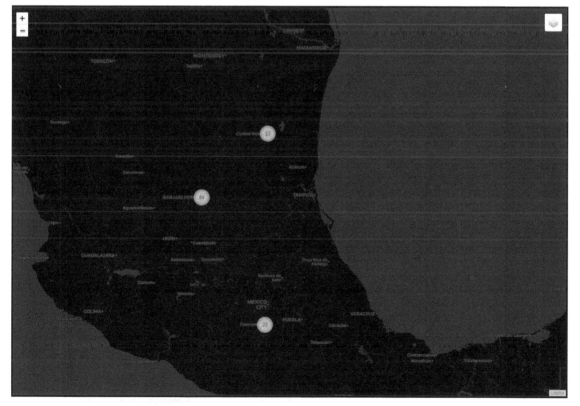

Cluster map of restaurants

Let's explore the data further and see how many unique cities are in the dataset:

```
print("Unique Cities in the dataset is
{}".format(geoplaces.city.nunique()))
```

The following output prints out that there are 17 cities in this dataset. There must be something wrong and if we dig deeper, we will see that there are duplicates because of mistyped names of cities in the dataset:

```
Unique Cities in the dataset is 17
```

So, let's summarize the unique cities within the dataset in `countplot`:

```
# Display cities in countplot
fig, ax = plt.subplots(figsize=(18,12))
sns.countplot(x="city",data=geoplaces, color="grey",
order=geoplaces['city'].value_counts().index, ax=ax)
plt.show()
```

As the following plot indicates, there are a lot of duplicated and mistyped city names in the dataset that need to be cleaned out. For example, **Cuernavaca** is misspelled as **cuernavaca**:

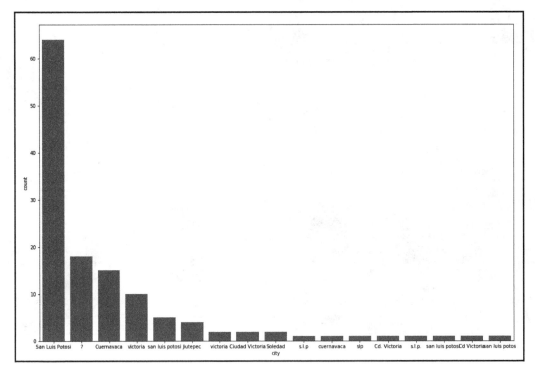

Cities countplot

Let's fix this and replace the mistakes with the correctly spelled city names. First, we group all of the cities with misspelled names like this:

```
cuer = ['Cuernavaca', 'cuernavaca', ]
slp = ['s.l.p.', 'San Luis Potosi', 'san luis potosi', 'slp', 'san luis
potos', 'san luis potosi ', 's.l.p']
ciudad = ['victoria ', 'victoria', 'Cd Victoria', 'Ciudad Victoria', 'Cd.
Victoria']
```

Then, replace the mistyped name with the correct one like this:

```
geoplaces['city']=geoplaces['city'].replace(slp,'San Luis Potosi' )
geoplaces['city']=geoplaces['city'].replace(ciudad,'Ciudad Victoria')
geoplaces['city']=geoplaces['city'].replace(cuer,'Cuernavaca')
```

If we redraw and look at the preceding `countplot` again, it looks like this:

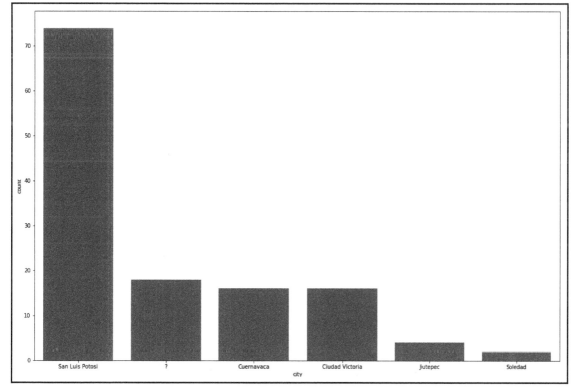

Cleaned cities countplot

This looks much cleaner. Let's explore some other features of this dataset. We will display a 2 x 2 plot of countplots for restaurant alcohol service, price, accessibility, and ambience:

```
# Subplots for four countplots
fig, ax = plt.subplots(2,2, figsize=(15,12))
sns.countplot(x="alcohol",data=geoplaces, color="grey", ax=ax[0][0])
ax[0][0].set_title('Alcohol Service')
sns.countplot(x="price",data=geoplaces, color="grey", ax=ax[0][1])
ax[0][1].set_title('Price Categories')
sns.countplot(x="accessibility",data=geoplaces, color="grey", ax=ax[1][0])
ax[1][0].set_title('Accessibility Categories')
sns.countplot(x="Rambience",data=geoplaces, color="grey", ax=ax[1][1])
ax[1][1].set_title('Rambience')
#fig.subplots_adjust(hspace=0.5)
plt.tight_layout()
plt.show()
```

The preceding code output is a 2 x 2 countplot. In the upper left-hand corner, we have an alcohol service countplot of restaurants. Most restaurants do not offer alcohol services, as you can see from the plot. Prices (the upper right-hand graph) also show restaurants with medium prices to have the highest counts. In the lower left-hand part of the plot, accessibility counts indicate that most restaurants do not have accessibility options. Finally, the ambience of restaurants (the lower right-hand graph) has two categories: familiar ambience as well as quiet ambience:

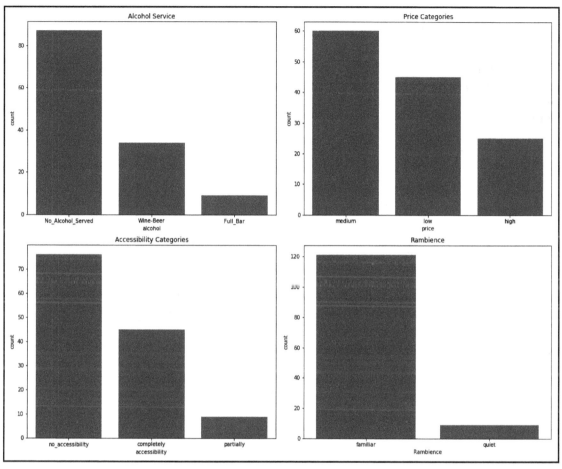

Countplots for alcohol service (upper-left), price (upper-right), accessibility (lower-left), and ambience (lower-right)

We are ready now to carry out the recommendation algorithm, but we first need to merge the dataset. Since both the `ratings` and `geoplaces` datasets have `placeID`, we can use this column to merge them both. We call our merged dataset simply `df`:

```
df = pd.merge(ratings, geoplaces, on='placeID')
```

Recommender systems

Recommender systems are one of the most commonly used practical systems in data science. In this section, we will focus on collaborative filtering, where the focus is on similarities between users. Depending on the past preference of users, this type of recommender system recommends items that users have liked or rated highly in the past. For this task, we will use `Surprise`, a Python scikit-learn library for building and analyzing recommender systems.

We first need to read the merged `df` into `Surprise`, set the rating scale of the dataset, and load data from `df` into `Surprise` data:

```
# Set rating scale of the dataset
reader = Reader(rating_scale=(0, 2))

# Load the dataframe with ratings.
data = Dataset.load_from_df(df[['userID', 'placeID', 'rating']], reader)
```

Now, we are set and can use the `Surprise` library functionalities. First, we will get a benchmark on this dataset from different available algorithms in `Surprise`. We will do cross-validation on the whole dataset and append the results in `bencmark_scores`. We also set random seed to `114` to get reproducible results:

```
benchmark_scores = []

random.seed(114)
np.random.seed(114)

# Iterate selected algorithms
for algorithm in [BaselineOnly(), SVD(), SVDpp(),
KNNWithMeans(),KNNWithZScore(), CoClustering(), NMF()]:
 # Cross-validation
 cv = cross_validate(algorithm, data, cv=5, verbose=False)

 # create df with cv results
 df_cv = pd.DataFrame.from_dict(cv).mean(axis=0)
 df_cv = df_cv.append(pd.Series([str(algorithm).split('
')[0].split('.')[-1]], index=['Algorithm']))
 benchmark_scores.append(df_cv)
```

Now, let's create a pandas `DataFrame` from the `benchmark_scores` list and see the results of each algorithm:

```
# Create results DataFrame from the benckmark scores
results =
pd.DataFrame(benchmark_scores).set_index('Algorithm').sort_values('test_rms
e')
results
```

`results` are shown as follows. Each algorithm and its result is displayed in `DataFrame`. In this particular dataset, the **SVDpp** algorithm, which is part of the matrix factorization recommender within the collaborative filtering algorithms, performs well. It has a **Root Mean Squared Error** (**RMSE**) of 0.65. The next best algorithm, in this case, happens to be the **KNNWithMeans** algorithm, directly derived from a basic nearest neighbors approach, with an RMSE of 0.66:

Algorithm	test_rmse	test_mae	fit_time	test_time
SVDpp	0.657078	0.549815	0.211962	0.006811
KNNWithMeans	0.661444	0.497731	0.003141	0.006092
SVD	0.666097	0.565743	0.062267	0.003131
KNNWithZScore	0.677907	0.502996	0.008874	0.006949
BaselineOnly	0.691530	0.595164	0.003867	0.001565
CoClustering	0.696392	0.516317	0.034050	0.001619
NMF	0.718128	0.549835	0.075321	0.001678

Algorithm benchmark results

You might get **fit_time** and **test_time** results that are different than the ones shown in our table. That is due to the computation power of the machines used.

We will apply these two algorithms, `SVDpp` and `KNNWithMeans`, into the dataset, and compare how they perform in different scenarios. First, let's set up cross-validation with 5 splits:

```
# define a cross-validation iterator
kf = KFold(n_splits=5)
```

Then, we will also set the two chosen algorithms, `SVDpp` and `KNNWithMeans`:

```
# Define algorithms
algo_knnwithMeans = KNNWithMeans()
algo_svdpp = SVDpp()
```

Let's start with the `KNNWithMeans` algorithm and apply it to our dataset.

KNNWithMeans

`KNNWithMeans` is a basic collaborative filtering algorithm, taking into account the mean ratings of each user. It is inspired directly by k-nearest neighborhood and the main tuning parameter is the maximum number of *k*. We will use the default, which is 40, but you can try and see how this changes the results.

First, we set a seed to make the results reproducible. Then, we loop through the data and split into training and test datasets according to the `kf` cross-validation we have created previously. Inside the loop, we first fit on the training dataset, then predict with the test dataset. We will calculate accuracy using RMSE metrics. Finally, we dump the data into a pandas `DataFrame` for later use:

```
random.seed(114)
np.random.seed(114)

for trainset, testset in kf.split(data):

    # train and test algorithm with KNNWithMeans.
    algo_knnwithMeans.fit(trainset)
    predictions_knnwithmeans = algo_knnwithMeans.test(testset)

    # Compute and print Root Mean Squared Error
    accuracy.rmse(predictions_knnwithmeans, verbose=True)
    dump.dump('./dump_KNNWithMeans', predictions_knnwithmeans,
algo_knnwithMeans)
```

The preceding fits on the training dataset, predicts on `testset`, and calculates `accuracy` based on RMSE metrics for each iteration. We can get the mean of RMSE for all iterations by averaging all RMSEs:

```
print("Average RMSE of the CV is: ",
np.mean(accuracy.rmse(predictions_knnwithmeans, verbose=False)))
```

We get this printed out from the preceding code:

```
Average RMSE of the CV is: 0.688647918600383
```

Now, let's load the dumped file of this algorithm and look closely at the predictions:

```
# Load the dump file
predictions_knnwithmeans, algo_knnwithMeans =
dump.load('./dump_KNNWithMeans')
```

After loading the dumped file, we can easily create a DataFrame from it like this and look at the first five rows:

```
df_knnithmeans = pd.DataFrame(predictions_knnwithmeans, columns=['uid',
'iid', 'rui', 'est', 'details'])
df_knnithmeans.head()
```

The output is pandas DataFrame and the following shows the first five rows of df_knnithmeans we have just created:

	uid	iid	rui	est	details
0	U1040	134999	1.0	0.983333	{'actual_k': 2, 'was_impossible': False}
1	U1032	135042	0.0	1.239111	{'actual_k': 12, 'was_impossible': False}
2	U1061	135058	1.0	1.420927	{'actual_k': 9, 'was_impossible': False}
3	U1118	134992	0.0	0.000000	{'actual_k': 2, 'was_impossible': False}
4	U1019	132856	0.0	0.000000	{'actual_k': 11, 'was_impossible': False}

df_knnithmeans head

This df_knnithmeans consists of five columns. The first one, uid, as you might recognize, is userID from our dataset; the second column, iid, is placeID of the restaurants. rui represents the rating of users with items. est is the estimated or predicted result from the algorithm. To calculate the error of each row, we can simply subtract the rui and est columns. Let's create a column called err to store the error results:

```
# Calculate the error
df_knnithmeans['err'] = abs(df_knnithmeans.est - df_knnithmeans.rui)
df_knnithmeans.head()
```

If we look at the first few rows here, you can see each row error from the `err` column we have just created:

	uid	iid	rui	est	details	err
0	U1040	134999	1.0	0.983333	{'actual_k': 2, 'was_impossible': False}	0.016667
1	U1032	135042	0.0	1.239111	{'actual_k': 12, 'was_impossible': False}	1.239111
2	U1061	135058	1.0	1.420927	{'actual_k': 9, 'was_impossible': False}	0.420927
3	U1118	134992	0.0	0.000000	{'actual_k': 2, 'was_impossible': False}	0.000000
4	U1019	132856	0.0	0.000000	{'actual_k': 11, 'was_impossible': False}	0.000000

df_knnithmeans head with error column

We will move to an `SVDpp` algorithm and later come back to compare the results of these two different algorithms.

SVDpp

The `SVDpp` algorithm is an extension of the SVD algorithm popularized by the fact that it won third place in the Netflix recommendation competition. `SVDpp` takes into account implicit rating, which is an improvement on the original SVD algorithm. We will carry out the same procedure as before but only change the algorithm from `KNNwithMeans` to `SVDpp`:

```
random.seed(114)
np.random.seed(114)

for trainset, testset in kf.split(data):
    # train and test algorithm with SVDpp.
    algo_svdpp.fit(trainset)
    predictions_svdpp = algo_svdpp.test(testset)
    # Compute and print Root Mean Squared Error
    accuracy.rmse(predictions_svdpp, verbose=True)
    # Dump the prediction into dataframe
    dump.dump('./dump_SVDpp', predictions_svdpp, algo_svdpp)
```

Let's print out the average RMSE of this algorithm:

```
print("Average RMSE of the CV is: ",
np.mean(accuracy.rmse(predictions_svdpp, verbose=False)))
```

The output indicates the following:

```
Average RMSE of the CV is: 0.667695280346118
```

This is an improvement from the KNNWithMeans result.

Let's also load the dumped file and create DataFrame with the results. We will also calculate the error of the predictions:

```
# Load the dump file
predictions_svdpp, algo_svdpp = dump.load('./dump_SVDpp')

df_svdpp = pd.DataFrame(predictions_svdpp, columns=['uid', 'iid', 'rui',
'est', 'details'])


df_svdpp['err'] = abs(df_svdpp.est - df_svdpp.rui)
df_svdpp.head()
```

This is the output of df_svdpp we have just created previously with the error rates:

	uid	iid	rui	est	details	err
0	U1040	134999	1.0	1.215454	{'was_impossible': False}	0.215454
1	U1032	135042	0.0	1.300336	{'was_impossible': False}	1.300336
2	U1061	135058	1.0	1.388325	{'was_impossible': False}	0.388325
3	U1118	134992	0.0	0.758690	{'was_impossible': False}	0.758690
4	U1019	132856	0.0	0.232720	{'was_impossible': False}	0.232720

df_svdpp head with error column

Although we can see that SVDpp performs better than KNNWithMeans in this case, we can compare these two algorithms to find out where each performs better than the other.

Comparison and interpretations

We can simply get the worst predictions of the algorithms by sorting them. Let's first get the worst predictions of df_knnithmeans:

```
df_knnithmeans.sort_values(by='err')[-10:]
```

As you can see from the following table, the worst prediction has an error of 2.00:

	uid	iid	rui	est	details	err
210	U1115	135054	0.0	1.644838	{'actual_k': 8, 'was_impossible': False}	1.644838
38	U1015	132866	0.0	1.711111	{'actual_k': 4, 'was_impossible': False}	1.711111
80	U1134	135026	0.0	1.711875	{'actual_k': 6, 'was_impossible': False}	1.711875
113	U1036	135065	0.0	1.757241	{'actual_k': 6, 'was_impossible': False}	1.757241
205	U1030	135016	0.0	1.833333	{'actual_k': 1, 'was_impossible': False}	1.833333
43	U1048	135048	0.0	1.839773	{'actual_k': 2, 'was_impossible': False}	1.839773
41	U1023	132733	2.0	0.151786	{'actual_k': 3, 'was_impossible': False}	1.848214
175	U1014	135082	0.0	1.884457	{'actual_k': 5, 'was_impossible': False}	1.884457
59	U1118	135021	2.0	0.000000	{'actual_k': 4, 'was_impossible': False}	2.000000
142	U1116	135059	0.0	2.000000	{'actual_k': 7, 'was_impossible': False}	2.000000

df_knnithmeans worst 10 predictions

We will do the same for `df_svdpp` to get the worst 10 predictions:

```
df_svdpp.sort_values(by='err')[-10:]
```

And here is the output of the worst 10 predictions for `df_svdpp`. Compared to the preceding table, you can see that this table has lower error rates. The worst prediction error is 1.79, compared to 2.00 from the preceding table:

	uid	iid	rui	est	details	err
194	U1027	135066	0.0	1.260576	{'was_impossible': False}	1.260576
1	U1032	135042	0.0	1.300336	{'was_impossible': False}	1.300336
97	U1005	135043	0.0	1.308369	{'was_impossible': False}	1.308369
175	U1014	135082	0.0	1.341835	{'was_impossible': False}	1.341835
210	U1115	135054	0.0	1.381919	{'was_impossible': False}	1.381919
43	U1048	135048	0.0	1.396731	{'was_impossible': False}	1.396731
38	U1015	132866	0.0	1.413020	{'was_impossible': False}	1.413020
80	U1134	135026	0.0	1.591442	{'was_impossible': False}	1.591442
113	U1036	135065	0.0	1.686855	{'was_impossible': False}	1.686855
142	U1116	135059	0.0	1.794109	{'was_impossible': False}	1.794109

df_svdpp worst 10 predictions

We can show the overall distribution of the prediction errors in `distplot` using `seaborn`. We will construct a 2 x 2 plot, where we will plot the prediction errors of `df_svdpp` and `df_knnithmeans`:

```
fig, ax = plt.subplots(1,2, figsize=(10,8))
sns.distplot(df_svdpp.err, ax=ax[0])
sns.distplot(df_knnithmeans.err, ax=ax[1])
ax[0].set_title('SVDpp')
ax[1].set_title('KNNwithMeans')
plt.show()
```

Here is the plot comparing the two algorithm errors:

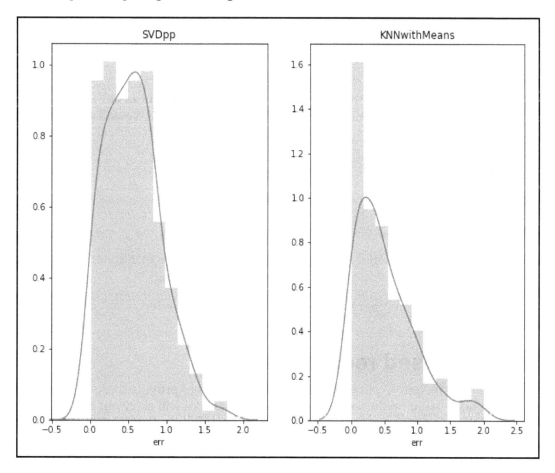

Comparison distplots

It seems both algorithms are right-skewed and have higher predictions around zero. So, what happens when a user has a smaller number of ratings? Remember that we had a mean rating of 1.19. Let's choose users with fewer than 5 ratings. This function is copied from the documentation of the `Surprise` library and calculates the number of items rated by a given user:

```
def get_Iu(uid):
    try:
        return len(trainset.ur[trainset.to_inner_uid(uid)])
    except ValueError: # user was not part of the trainset
        return 0
```

We will use this function to calculate the user rating for both `df_knnithmeans` and `df_svdpp`:

```
df_knnithmeans['Iu'] = df_knnithmeans.uid.apply(get_Iu)
df_svdpp['Iu'] = df_svdpp.uid.apply(get_Iu)
```

We can now compare the error rates of the two `DataFrame` when users only have fewer than 5 ratings and see which one of the algorithms performs better. We will use the mean here:

```
df_knnithmeans[df_knnithmeans.Iu < 5].err.mean(),
df_svdpp[df_svdpp.Iu < 5].err.mean()
```

The output is as follows:

```
(0.5104404961856474, 0.580852352863545)
```

This is for `df_knnithmeans` and `df_svdpp` respectively. It seems that `KNNwithMeans` is performing much better than `SVDpp` when users have fewer ratings.

In the next section, we will cover **location-based** (**LB**) recommenders.

Location-based recommenders

LB recommenders include explication location components to provide more relevant recommendations based on the location of users or items. We will carry out a simple LB recommendation using the k-means clustering techniques we covered in Chapter 4, *Making Sense of Humongous Location Datasets*.

We will first fit on a small sample of `df` and get the labels. Let's also print out the number of `k` and `silhouette_score`, which is the metric we will use and the number of clusters:

```
kmeans = KMeans(n_clusters=24, init='k-means++')
X_sample = df[['longitude','latitude']].sample(frac=0.1)
kmeans.fit(X_sample)
y = kmeans.labels_

print("k = 24", " silhouette_score ", silhouette_score(X_sample, y,
metric='euclidean'))
```

Our `k` value is equal to `k = 24` and `silhouette_score`, which is `0.5461956922155007`.

Now, we will predict on the whole dataset and populate in a new column in our `df['cluster']`, let's also get a sample of 10 rows from `df`:

```
df['cluster'] = kmeans.predict(df[['longitude','latitude']])
df[['userID','latitude','longitude','placeID','cluster']].sample(10)
```

Here is the output of the sample table in the preceding code. So, as you can see from the following table, we have a new `cluster` column where each coordinate point is in a cluster:

	userID	latitude	longitude	placeID	cluster
96	U1022	22.156883	-100.978485	135060	21
802	U1005	22.151060	-100.977659	135041	3
109	U1098	22.156883	-100.978485	135060	21
669	U1126	22.140129	-100.944872	135069	8
859	U1009	22.145108	-100.989547	135059	12
484	U1065	23.726819	-99.126506	132561	10
228	U1011	23.730925	-99.145185	132564	23
645	U1056	22.153703	-100.979033	135062	21
768	U1019	22.150849	-100.939752	132830	8
1053	U1052	22.141238	-100.923925	132869	4

Sample of df with clusters

Note that the values in the preceding table might change because we are using a random sample.

We will use top-rated venues for our recommendation. This can be changed to, for example, the number of rating per restaurant, or more tailored recommenders such as different cuisines of restaurants. Let's get the highest rated restaurants. Here, we will not only include `rating` but also `food_rating` and `service_rating` since we have many restaurants with a rating of 2. We call this `topvenues_df`:

```
topvenues_df = df.sort_values(by=['rating', 'food_rating',
'service_rating'], ascending=False)
```

We create a simple function that receives `df`, as well as `latitude` and `longitude`. The function first predicts the cluster of coordinates provided. Once it gets the cluster number, it passes through `df` provided in this `topvenues_df` and gets the first top-rated name of the restaurant in this cluster. Finally, the function prints out recommendations:

```
def recommend_restaurants(df, longitude, latitude):
    # Predict the cluster for longitude and latitude provided
    cluster =
kmeans.predict(np.array([longitude,latitude]).reshape(1,-1))[0]
    # Get the best restaurant in this cluster
    name = df[df['cluster']==cluster].iloc[0]['name']
    print('"{}" is recommended'.format(name))
```

Let's also create a function that plots a Folium map for both user location and restaurant locations. This function takes `df`, a user coordinates, and the restaurant name produced by the preceding function:

```
def create_folium_map(df, user_coords, restaurant_name):
    m = folium.Map(
    location=user_coords,
    zoom_start=10,
    tiles='Stamen Terrain'
    )
    folium.Marker(
    location=user_coords,
    popup='User Location',
    icon=folium.Icon(icon='cloud')
    ).add_to(m)
    folium.Marker(
    location=list(df[df['name'] == restaurant_name][['latitude',
'longitude']].iloc[0]),
    popup='Restaurant Location',
```

```
icon=folium.Icon(color='red',icon='info-sign')
).add_to(m)
return m
```

Let's use the `recommend_restaurants` function to recommend restaurants:

```
recommend_restaurants(topvenues_df,-99.145185, 23.730925)
```

The output prints out the recommendations:

```
"TACOS EL GUERO" is recommended
```

Here is another example with different locations:

```
recommend_restaurants(topvenues_df, -100.939752, 22.150849)
```

This one prints out another restaurant name:

```
"Rincon Huasteco" is recommended
```

We utilize the `create_folium_map` function to display user location and restaurant location in a map for the last example, which has recommended "`Rincon Huasteco`". Let's first create variables that hold `user_coords` and `restaurant_name`:

```
user_coords = [22.120849, -100.839752]
restaurant_name = "Rincon Huasteco"
```

Now, let's pass these variables to the `create_folium_map` function:

```
create_folium_map(df, user_coords, restaurant_name)
```

The output is a map, as follows, where the cloud icon is the user location and the information sign icon is the restaurant location:

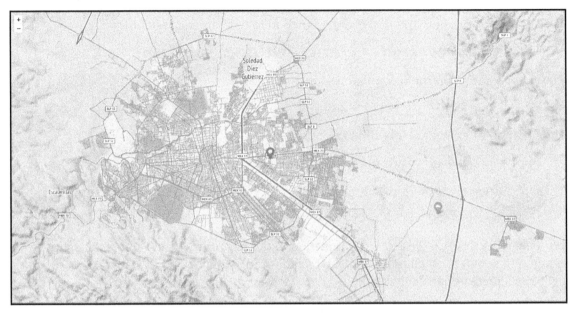

Recommendation map

Congratulations! This is just the beginning of your journey in geospatial data science. While reading this book, you have been introduced to a broad and essential range of geospatial Python libraries, as well as real-world applications. I hope that this will be your inspiration to continue learning and working on the vast array of geospatial data science projects out there.

Summary

In this chapter, we explored different types of recommenders. In the first section, we did an exploratory data analysis to get a grasp of the dataset. We also preprocessed and cleaned a dataset, as well as merged different DataFrames. In the second section, we learned about collaborative filtering recommenders and used two different algorithms. In the final section, we explored LB recommenders and used the k-means clustering algorithm and top-rated restaurants to recommend restaurant venues based on the location of the user.

Other Books You May Enjoy

If you enjoyed this book, you may be interested in these other books by Packt:

Hands-On Geospatial Analysis with R and QGIS

Shammunul Islam

ISBN: 978-1-78899-167-4

- Install R and QGIS
- Get familiar with the basics of R programming and QGIS
- Visualize quantitative and qualitative data to create maps
- Find out the basics of raster data and how to use them in R and QGIS
- Perform geoprocessing tasks and automate them using the graphical modeler of QGIS

Mastering Geospatial Development with QGIS 3.x - Third Edition

Dr. John Van Hoesen, GISP, Luigi Pirelli, Et al

ISBN: 978-1-78899-989-2

- Create and manage a spatial database
- Get to know advanced techniques to style GIS data
- Prepare both vector and raster data for processing
- Add heat maps, live layer effects, and labels to your maps
- Master LAStools and GRASS integration with the Processing Toolbox

Leave a review - let other readers know what you think

Please share your thoughts on this book with others by leaving a review on the site that you bought it from. If you purchased the book from Amazon, please leave us an honest review on this book's Amazon page. This is vital so that other potential readers can see and use your unbiased opinion to make purchasing decisions, we can understand what our customers think about our products, and our authors can see your feedback on the title that they have worked with Packt to create. It will only take a few minutes of your time, but is valuable to other potential customers, our authors, and Packt. Thank you!

Index

map projections comparison
  reference 48
model
  building 36
  data, validating 36, 38, 39
  error metrics, validating 36, 38, 39
Modifiable Area Unit Problem (MAUP) 10

N

noise 66

O

Open Street Maps (OSM) 113, 114

P

polygon geometries 90
probabilistic 10
projections 48

R

raster data 10
recommender systems
  about 140, 141, 142
  comparison 145, 147, 148
  interpretations 145, 147, 148
  KNNWithMeans algorithm 142, 143, 144
  location-based (LB) 148, 150, 151, 152
  SVDpp algorithm 144, 145
Relational Database Management System
    (RDBMS) 9
road data
  exploring 114, 115
road network graph
  shortest path, analyzing 124

road network
  graph, creating on 113
Root Mean Squared Error (RMSE) 141
Root Mean Squared Logarithmic Error (RMSLE)
    25

S

spatial autocorrelation 70
  point, in polygon 72, 73
spatial data
  processing 27
spatial joins 52, 55
spatial operations
  about 48
  buffer analysis 49, 50
  projections 48
  spatial joins 52, 55
stochastic 10
SVDpp algorithm 144, 145

T

taxi zones
  in New York 27
  staxi zones 31
  visualization 27, 28
Tensor Processing Unit (TPU) 17
the small-world phenomenon 104
time values
  as features 26
Tobler's first law 9
topology 86, 87, 91, 93

V

vector data 10